The Darkling Beetles of the Sinai Peninsula

An expanded and updated edition of the out-of-print 2003 supplementum of *Zoology in the Middle East*, this concise guide to *Darkling Beetles of the Sinai Peninsula* has been sought after by researchers in taxonomy, faunistics and biogeography. The new book includes two additional subfamilies of tenebrionid beetles (4–5 species), identification keys and more than 90 colour photographs and species distribution maps.

Zoogeographically speaking, the Sinai Peninsula is a crossroad and, at the same time, a center of speciation. Despite its generally arid character, the region harbours a wide range of habitats, from sea level to over 2,500 m above. About 10 percent of the Sinai darkling beetles are endemic to the area. The inclusion of species photographs and identification keys makes this book an invaluable reference field guide, for both specialists and non-specialists, who will thus be able to discover the taxonomic and phylogenetic diversity of darkling beetles in the Sinai Peninsula.

The Darkling Beetles of the Sinai Peninsula
Coleoptera: Tenebrionidae

Martin Lillig
Tomáš Pavlíček

CRC Press
Taylor & Francis Group
Boca Raton London New York

CRC Press is an imprint of the
Taylor & Francis Group, an **informa** business

First edition published 2023
by CRC Press
6000 Broken Sound Parkway NW, Suite 300, Boca Raton, FL 33487-2742

and by CRC Press
4 Park Square, Milton Park, Abingdon, Oxon, OX14 4RN

CRC Press is an imprint of Taylor & Francis Group, LLC

© 2023 Martin Lillig and Tomáš Pavlíček

Reasonable efforts have been made to publish reliable data and information, but the author and publisher cannot assume responsibility for the validity of all materials or the consequences of their use. The authors and publishers have attempted to trace the copyright holders of all material reproduced in this publication and apologize to copyright holders if permission to publish in this form has not been obtained. If any copyright material has not been acknowledged please write and let us know so we may rectify in any future reprint.

Except as permitted under U.S. Copyright Law, no part of this book may be reprinted, reproduced, transmitted or utilized in any form by any electronic, mechanical, or other means, now known or hereafter invented, including photocopying, microfilming, and recording, or in any information storage or retrieval system, without written permission from the publishers.

For permission to photocopy or use material electronically from this work, access www.copyright.com or contact the Copyright Clearance Center, Inc. (CCC), 222 Rosewood Drive, Danvers, MA 01923, 978-750-8400. For works that are not available on CCC please contact mpkbookspermissions@tandf.co.uk

Trademark notice: Product or corporate names may be trademarks or registered trademarks and are used only for identification and explanation without intent to infringe.

ISBN: 978-1-032-18720-4 (hbk)
ISBN: 978-1-032-18721-1 (pbk)
ISBN: 978-1-003-25590-1 (ebk)

DOI: 10.1201/9781003255901

Typeset in Times
by KnowledgeWorks Global Ltd.

Dedication

I owe a great debt of gratitude to the people who inspired me to pursue coleopterology, namely my late parents who encouraged my interests in entomology, Prof. (Em.) Dr. Peter Nagel who led me to the beetles and especially to the Tenebrionidae, the many colleagues with whom I have friendly and constructive exchanges for a long time and my loved ones who have patiently suffered and will continue to suffer from the many hours I have spent and will continue to spend on coleopterology.

ML

To my parents who kindled my interest in science and kept faith in my academic and scientific progress, to my wife Patricia Cardet whom apart from supporting my scientific career also accompanies me in many collection trips and to Dr Ottò Merkl (Hungarian Natural History Museum, Budapest) for his inspiration in the study of darkling beetles.

TP

Contents

Foreword xi
Preface xiii
Authors xv
List of Acronyms and Abbreviations xvii
Introduction xix

Chapter I Description of the Sinai Peninsula 1

          I.1   Physical Borders and Landscapes 1
          I.2   Climate and Paleoenvironment 3
          I.3   Flora and Fauna 4
                  I.3.1   Mosses and Vascular Plants 4
                  I.3.2   Vertebrates 5
                  I.3.3   Invertebrates 5
          I.4   Local Geographic Names 6

Chapter II Collection Methods and Analysis of the Darkling Beetles Distribution 9

Chapter III Species Richness and Endemism of Tenebrionidae Beetles in the Sinai Peninsula 11

          III.1   Global Distribution Patterns of the Tenebrionidae in the Sinai Peninsula 11

Chapter IV Ecology of Desert Darkling Beetles 17

Chapter V Identification Keys for the Tenebrionidae Taxa Recorded on the Sinai Peninsula 19

          Key 1 Identification Key to the Subfamily Level 19
          Key 2 Identification Key for the Subfamily Pimeliinae 20
          Key 3 Identification Key for the Subfamilies Blaptinae and Tenebrioninae 29
          Key 4 Identification Key for the Subfamily Diaperinae 32
          Key 5 Identification Key for the Subfamily Alleculinae 33

Chapter VI Darkling Beetles Present in the Sinai Peninsula 35

          VI.1   Subfamily Pimeliinae Latreille, 1802 35
                  VI.1.1   Tribe Adelostomini Solier, 1834 35
                            VI.1.1.1   Genus *Adelostoma* Duponchel, 1827 35

		VI.1.1.2 Genus *Machlopsis* Pomel, 1871............36

- VI.1.2 Tribe Adesmiini Lacordaire, 1859........................36
  - VI.1.2.1 Genus *Adesmia* Fischer von Waldheim, 1822.....................36
- VI.1.3 Tribe Akidini Billberg, 1820................................44
  - VI.1.3.1 Genus *Akis* Herbst, 1799......................44
- VI.1.4 Tribe Erodiini Billberg, 1820................................45
  - VI.1.4.1 Genus *Amnodeis* Miller, 1858..............45
  - VI.1.4.2 Genus *Erodius* Fabricius, 1775............45
- VI.1.5 Tribe Pimeliini Latreille, 1802...........................49
  - VI.1.5.1 Genus *Astorthocnemis* Lillig & Pavlíček, 2002......................................49
  - VI.1.5.2 Genus *Pimelia* Fabricius, 1775...........49
  - VI.1.5.3 Genus *Prionotheca* Dejean, 1834........52
  - VI.1.5.4 Genus *Pterolasia* Solier, 1836............52
  - VI.1.5.5 Genus *Thriptera* Solier, 1836...............52
  - VI.1.5.6 Genus *Trachyderma* Latreille, 1828.....53
- VI.1.6 Tribe Sepidiini Eschscholtz, 1829.......................55
  - VI.1.6.1 Genus *Sepidium* Fabricius, 1775..........55
  - VI.1.6.2 Genus *Vieta* Laporte de Castelnau, 1840..................................56
- VI.1.7 Tribe Stenosini Schaum, 1859 (1934)..................56
  - VI.1.7.1 Genus *Dichillus* Jacquelin du Val, 1860..56
  - VI.1.7.2 Genus *Microtelus* Solier, 1838..............56
  - VI.1.7.3 Genus *Mitotagenia* Reitter, 1916.........56
  - VI.1.7.4 Genus *Stenosis* Herbst, 1799................57
- VI.1.8 Tribe Tentyriini Eschscholtz, 1831......................57
  - VI.1.8.1 Genus *Cyphostethe* Marseul, 1867.......57
  - VI.1.8.2 Genus *Dailognatha* Steven, 1828.........57
  - VI.1.8.3 Genus *Hegeterocara* Reitter, 1900.......57
  - VI.1.8.4 Genus *Hionthis* Miller, 1861................58
  - VI.1.8.5 Genus *Imatismus* Dejean 1834............58
  - VI.1.8.6 Genus *Mesostena* Eschscholtz, 1831.....58
  - VI.1.8.7 Genus *Micipsa* Lucas, 1855..................61
  - VI.1.8.8 Genus *Oterophloeus* Desbrochers des Loges, 1881....................................62
  - VI.1.8.9 Genus *Oxycara* Solier, 1835.................62
  - VI.1.8.10 Genus *Prochoma* Solier, 1835............63
  - VI.1.8.11 Genus *Scelosodis* Solier, 1835............65
  - VI.1.8.12 Genus *Schweinfurthia* Andres, 1922....65
  - VI.1.8.13 Genus *Tentyria* Latreille, 1802............65
  - VI.1.8.14 Genus *Tentyrina* Reitter, 1900.............66
  - VI.1.8.15 Genus *Thraustocolus* Kraatz, 1865......67
  - VI.1.8.16 Genus *Trichosphaena* Reitter, 1916....67

Contents

	VI.1.9	Tribe Zophosini Solier, 1834 67
		VI.1.9.1 Genus *Zophosis* Latreille, 1802 67
VI.2	Subfamily Blaptinae Leach, 1815 69
	VI.2.1	Tribe Blaptini Leach, 1815 69
		VI.2.1.1 Genus *Blaps* Fabricius, 1775 69
	VI.2.2	Tribe Dendarini Mulsant & Rey, 1854 71
		VI.2.2.1 Genus *Dendarus* Dejean, 1821 71
	VI.2.3	Tribe Opatrini Brullé, 1832 71
		VI.2.3.1 Genus *Ammobius* Guérin-Méneville, 1844 71
		VI.2.3.2 Genus *Clitobius* Mulsant & Rey, 1859 71
		VI.2.3.3 Genus *Proscheimus* Desbrochers des Loges, 1881 72
		VI.2.3.4 Genus *Diphyrrhynchus* Fairmaire, 1849 72
		VI.2.3.5 Genus *Gonocephalum* Solier, 1834 72
		VI.2.3.6 Genus *Mesomorphus* Miedel, 1880 74
		VI.2.3.7 Genus *Opatroides* Brullé, 1832 74
		VI.2.3.8 Genus *Scleropatroides* Löbl & Merkl, 2003 75
		VI.2.3.9 Genus *Eurycaulus* Fairmaire, 1868 75
		VI.2.3.10 Genus *Sclerum* Dejean, 1834 75
	VI.2.4	Tribe Pedinini Eschscholtz, 1829 77
		VI.2.4.1 Genus *Drosochrus* Erichson, 1843 77
		VI.2.4.2 Genus *Leichenum* Dejean, 1834 77
VI.3	Subfamily Tenebrioninae Latreille, 1802 77
	VI.3.1	Tribe Alphitobiini Reitter, 1917 77
		VI.3.1.1 Genus *Alphitobius* Stephens, 1829 77
	VI.3.2	Tribe Helopini Latreille, 1802 78
		VI.3.2.1 Genus *Hedyphanes* Fischer von Waldheim, 1820 78
	VI.3.3	Tribe Melanimonini Seidlitz, 1894 78
		VI.3.3.1 Genus *Cheirodes* Gené, 1839 78
	VI.3.4	Tribe Scaurini Billberg, 1820 79
		VI.3.4.1 Genus *Scaurus* Fabricius, 1775 79
	VI.3.5	Tribe Triboliini Gistel, 1848 80
		VI.3.5.1 Genus *Tribolium* MacLeay, 1825 80
VI.4	Subfamily Diaperinae Latreille, 1802 80
	VI.4.1	Tribe Crypticini, Brullé, 1832 80
		VI.4.1.1 Genus *Pseudoseriscius* Español, 1950 80
	VI.4.2	Tribe Phaleriini Blanchard, 1845 81
		VI.4.2.1 Genus *Phaleria* Latreille, 1802 81
		VI.4.2.2 Genus *Phtora* Germar, 1836 81

	VI.4.3	Tribe Trachyscelini Blanchard, 1845 82
		VI.4.3.1 Genus *Trachyscelis* Latreille, 1809 82

VI.5 Subfamily Alleculinae Laporte de Castelnau, 1840 82
    VI.5.1 Tribe Alleculini Laporte de Castelnau, 1840 82
        VI.5.1.1   Genus *Mycetocharina* Seidlitz, 1890 82
    VI.5.2 Tribe Cteniopodini, Seidlitz, 1896 82
        VI.5.2.1   Genus *Cteniopus* Solier, 1835 82
        VI.5.2.2   Genus *Omophlus* Dejean, 1834 82

Chapter VII   Incorrect and Doubtful Records (In Alphabetical Order) 107

Chapter VIII   Species Expected to Be Found on the Sinai Peninsula (In Alphabetical Order) 117

Bibliography 121

Taxonomic Index 141

Foreword

Darkling beetles are among the most conspicuous animals of deserts and semi-deserts. This is true at least from an insect maniac's perspective who surveys the soil surface not only to avoid snake bites or stings of scorpions but also to discover and enjoy the diversity of desert arthropods. The ecological importance of tenebrionids is evident from their sheer abundance, which is sometimes visible at daytime but even more so at night. Therefore, when exploring arid regions, the attention of even the most orthodox carabidologists may be diverted to darkling beetles – at least temporarily. This happened to me during my first expeditions to the Sahara Desert.

In order to cope with high surface temperatures, tenebrionids has evolved a wide range of leg structures, comprising sand shoes as well as stilts. In the morning, the sand is often covered with numerous tracks of different structures – transient traces of the night activity of different species. It is an impressive experience to realise the extraordinary diversity in shape and structure of tenebrionids, which in turn is an expression of their diverse modes of life and adaptations to various habitats, including the most inhospitable ones.

The readership of this book is unlikely to include people who know darkling beetles at best as mealworms fed to their bird or reptile pets, or as pests of stored food such as the *Tribolium* flour beetle. Most will be specialists, and this more professional readership will greatly appreciate the taxonomic data and information on synonyms carefully and reliably compiled in this book. In the first place, the book provides an excellent overview of the faunistics of a zoogeographically very important area and is, therefore, a valuable source of information for biogeographers. In addition, it is an important reference for all persons interested in the region and the conservation of its particular biodiversity. Both authors visited the area several times to collect beetles, discovered species whose presence on the Sinai Peninsula was not previously known, rediscovered rare species, and even new species for science. Their success was founded on a decent knowledge of behavioural and life-history traits of tenebrionids. By deriving specific distribution patterns within the Sinai Peninsula, the authors contribute significantly to our knowledge of the overall ranges. A considerable amount of material had already been collected earlier, however, most originated from localities along the main routes, supplemented with a few casual collections elsewhere.

Zoogeographically, the Sinai Peninsula represents a crossroad and contact zone for faunas of different origins and, at the same time, a center of speciation. Despite its generally arid character, the region harbours a wide range of habitats, ranging from sea level to more than 2,500 m altitude. The presence of endemic taxa in the mountains in the southern part of the peninsula mirrors climatic and vegetational fluctuations during the recent earth history. About 10 percent of all Sinai darkling beetles are endemic to the area.

As long as accurate taxonomic and faunistic data are missing, discussions about biodiversity conservation remains pure academic exercises. At worst, well-intended conservation efforts may be compromised by ill-founded data. Profound taxonomic knowledge is therefore a prerequisite for all faunistic studies. These are in turn the basis for spatial analyses of biodiversity. Extending further on this chain of argument, we have to realise that scientific collections are custodians of the foundations of our knowledge of biodiversity. The physical basis of this knowledge comprises specimens and records of their origin, habitat and collection details. Without specialists capable of critically evaluating these data, this information would not be accessible to the scientific community, and surely not to the public interested in conservation.

Fortunately, idealists are keeping track. Led by their enthusiasm for a particular taxon, they mature to specialists providing invaluable contributions not only to the advances of science but also to the protection of biodiversity. We have to thank Martin Lillig and Tomáš Pavlíček for taking this task with enthusiasm and care. In numerous publications, they have already given proof of their scientific expertise. This book will be a significant addition to their personal record and an important and lasting contribution to science. I am quite confident that many users with different scientific and geographical backgrounds will profit from this publication. I trust it will find the wide audience it deserves.

Peter Nagel
Professor (Emeritus) for Environmental Sciences/Biogeography
University of Basel, Switzerland

Preface

It is now almost 20 years since *The Darkling Beetles of the Sinai Peninsula* was first released. In the meantime, we hardly got to see any more material from the Sinai. Apparently, not many darkling beetles have been collected on the peninsula for the past two decades. At least, we did not learn of new larger collections. So why a new edition?

One reason is that we have received repeated enquiries as to whether a copy of the book is still available for purchase. Unfortunately, not. The book has been sold out for years now. This alone would be a good reason for a new edition. Another reason is that many readers (rightly) missed an identification key and illustrations of the beetles in the first edition. We wanted to make up for this omission. In this now expanded version, we take the opportunity to correct errors found in the first edition and to apply new results of research on the Tenebrionidae to the darkling beetles of the Sinai Peninsula. For the systematics of the darkling beetles, new insights have been gained, mainly by molecular genetic methods. A new subfamily has been proposed, new species described, new synonymies recognised.

All these reasons motivated us to completely revise the book. We hope to provide a useful compendium for all readers interested in nature, especially in the Tenebrionidae of the Sinai Peninsula.

Martin Lillig and Tomáš Pavlíček

Authors

Martin Lillig, PhD, works for the environmental and nature conservation association *Bund für Umwelt und Naturschutz Deutschland* (Friends of the Earth, Germany) and freelances in nature conservation. He received his doctorate in environmental sciences from the Faculty of Science at the University of Basel, Switzerland. Since studying biogeography in the 1980s at the University of Saarland, Germany, he has been particularly interested in entomology, especially beetles. The taxonomy, systematics and zoogeography of the Tenebrionidae are his special fields. He has published about 70 scientific papers in peer-reviewed journals and several book chapters. Since 2020, he has been subject editor of the journal *Zootaxa* for Tenebrionidae. He also writes peer reviews for numerous other journals. On his research trips he visited Zimbabwe, Algeria, Tunisia, Egypt (Sinai Peninsula), Israel, Jordan, the Canary Islands, Madeira, Cape Verde Islands and various European countries like Portugal, Malta, Greece, and so on.

Tomáš Pavlíček, RNDr, PhD, is a senior researcher at the Institute of Evolution, University of Haifa, Israel. He obtained a PhD (= CSc.) from the Institute of Entomology, České Budějovice, Czech Academy of Sciences. He has been conducting parallel research in entomology, vermicology and botany (cereals), always interested in the fields of evolutionary genetics, population genetics, ecology and zoogeography. He has co-authored eight monographs and published more than 130 scientific papers in peer-reviewed journals. His research trips have taken him to regions of the world as varied as French Guiana, Brazil, New Caledonia, Mayotte, Guadeloupe, Mauritius, Japan, Turkey, Portugal, the Caucasus, Jordan, Israel and the Sinai Peninsula.

List of Acronyms and Abbreviations

a.s.l.	above sea level
auct.	used in the sense of subsequent author(s) with the meaning that the species name is not the original one(s) – derived from the Latin word *auctorum* for "of authors"
auct. nec	author names used with *nec* indicate a misapplied name
BGSS	Biogeographische Sammlung der Universität des Saarlandes, Saarbrücken, now Zentrum für Biodokumentation, Landsweiler-Reden
BM	Natural History Museum, London
CGW	Collection of Cilli and Gerhard Wagner, Hamburg
det.	the name of a person who identified a species, subspecies or variant under consideration
E, N, S, W	east, north, south, west
GF	Collection of Georg Frey in the Naturhistorische Museum, Basel
HJB	Collection of Hans J. Bremer, Melle, now in ZSM
in litt.	*in litteris* = in correspondence
L.P.	Leg. M., M., S. Lillig and T. Pavlíček
leg.	from Latin *legit* he, she or it collected
MCSNM	Museo Civico di Storia natural di Milano
ML	Collection of Martin Lillig, Saarbrücken
MNHNP	Muséum national d'Histoire naturelle, Paris
MNHUB	Museum für Naturkunde der Humboldt-Universität, Berlin
MZH	Zoologiska Museet Universitet, Helsinki
nov. gen.	*genus novum*
n. sp., n. spp.	"species nova" – new species (n.sp.: singular, s.spp: plural)
n. syn.	new synonym
nec	"and not", from Latin *nec mole sapere*
NHMW	Naturhistorisches Museum Wien
NMP	Národní Muzeum, Praha
No. (= Nr.)	number (singular, plural is Nos. or Nrs.)
nom. n.	*nomen nudum* (singular, *nomina nuda* plural), a "naked", unavailable name that was not supported by the species description as requested
ONHM	Oman Natural History Museum, Muscat
P.	Leg. T. Pavlíček (Do not confuse P. with P [see below])
P	statistical probability
p.	page (singular, plural is abbreviated as pp.)
pr.	prope (lat.) = near
PT	paratype(s)

s. str.	*sensu stricto* (in the strict sense)
s. l.	*sensu lato* (in the broad sense)
SB	Collection of Stanislav Bečvář, České Budějovice
SMNS	Staatliches Museum für Naturkunde Stuttgart
ssp.	subspecies
ssp. n.	subspecies nova
syn. restit.	restored synonym
t.	teste (testified by)
TAU	Steinhardt Museum of Natural History at Tel Aviv University
TP	Collection of Tomáš Pavlíček, Haifa
var.	varietas (variety)
var. n.	varietas nova (new variety)
vid.	videbatur (lat.) = seen (by)
ZSM	Zoologische Staatssammlung München

Introduction

The Sinai Peninsula has played an important role in human history and its influence often goes far beyond the East Mediterranean region. Nevertheless, large "white spots" remain in our knowledge of the insect diversity of the peninsula. As a contribution to the knowledge of regional biodiversity, taxonomy and zoogeography of beetles, we decided to revise and update the available information about darkling beetles (Tenebrionidae including the six following subfamilies: Lagriinae, Pimeliinae, Blaptinae, Tenebrioninae, Diaperinae and Alleculinae) of the Sinai Peninsula. Data provided for all species includes the quotations of the original descriptions, synonyms, list of localities where the species were collected in the Sinai Peninsula, their general distribution and, where appropriate, comments on their taxonomy and ecology. Distribution maps are added for seven significantly differing species, in their prevalence, between the south and the north Sinai and photos of the 92 taxa out of 127 species and two subspecies of *Adesmia metallica*.

The darkling beetles studied here are part of the authors' own collections M., M., S. Lillig and T. Pavlíček XI.-XII.1992, T. Pavlíček IV.1993, IV.1995, southern part of the Sinai Peninsula), as well as of collections of N. Rech (III.1996, III.1997, Nuweiba), W. Schawaller (XII.1997-I.1998, south-eastern Sinai) and W. G. Ullrich (IV. 1992, X.1992, IV.1993, southern Sinai). Other specimens, in particular the Erodiini and Adesmiini, are part of the following collections: The Natural History Museum in London, Collection of G. Frey in Naturhistorisches Museum in Basel, Museo Civico di Storia Naturale di Milano, Museum für Naturkunde der Humboldt Universität in Berlin, Staatliches Museum für Naturkunde in Stuttgart, Muséum national d'Histoire naturelle in Paris, Naturhistorisches Museum Wien, Ditsong National Museum of Natural History (formerly Transvaal Museum) in Pretoria, Természettudományi Muzeum in Budapest and Zoologiska Museet Universitet in Helsinki. Some references are made to the entomological collection of the Tel Aviv University (see "List of Acronyms and Abbreviations").

To update the taxonomy and distribution of a large family of beetles such as Tenebrionidae, in a territory of about 61,000 km^2 like the Sinai Peninsula, is a difficult task. In our effort to discuss and evaluate not only new data but also published ones on the darkling beetles of this part of the world, we might have, not purposely, omitted some publications. The preparation of this book would not have been possible without the precious help of many of our colleagues and friends. We wish to thank Dr M. Baehr† (Munich), Dr J. Beard (London), Prof Dr H. J. Bremer (Osnabrück), Dr S. Endrödy-Younga† (Pretoria), Ing. S. Bečvář (České Budějovice), Prof A. Freidberg† (Tel Aviv), Dr M.D. Gallagher† (Muscat), Dr C. Girard (Paris), Dr D. Kempster (London), Dr C. Leonardi (Milano), Dr O. Merkl† (Budapest), Ms Dr N. Rech (Deming), Dr W. Schawaller (Stuttgart), Dr H. Schönmann† (Vienna), Dr H. Silfverberg (Helsinki), Dr M. Uhlig (Berlin), Ms C. and Mr G. Wagner (Hamburg) for lending us specimens from their own collections and/or collections of which they are curators. We are grateful to Dr M.-L. Penrith (Pretoria) for identification of the species of the genus *Zophosis* and to J. Ferrer† for the identification of the genus *Gonocephalum* in M. Lillig's collection. We are also much indebted to Ms A. Cieslak née Zahm (Tholey) for providing us with valuable literature, and to Ms P. Cardet (Haifa) for lexical and stylistic comments.

I Description of the Sinai Peninsula

I.1 PHYSICAL BORDERS AND LANDSCAPES

The Sinai Peninsula (in Arabic *Shibh Jazirat Sina* or شبه جزيرة سيناء) is a geologically, climatically and geographically highly variable triangular plateau separating southwest Asia from north-east Africa (Figure I.1). The Sinai Peninsula can be regarded as one of the most important past and present geographic and cultural bridges between both continents. The northern and southern borders of the peninsula are established by the Mediterranean Sea and the Red Sea (Figure I.1). However, the eastern border with Israel is an arbitrary line (Figure I.1). From the ethnographic, geologic and climatological point of view, the Negev Desert and the western Sinai are a single geographic unit (Tsoar 1995).

The separation of the peninsula from Africa through the Gulf of Suez Rift took place during the Miocene and from Arabia through the Araba-Dead Sea Transform fault during the Late Miocene/Pliocene (Kuss & Bachmann 1996). The shallow Gulf of Suez Rift Valley was partly filled up with sea water (Gulf of Suez and Bitter Lakes) and throughout the rest of the rift (Isthmus of Suez) artificial canals were built (Figure I.1). The first canal was dug during the pharaonic period of Egypt and the second, the Suez Canal (Figure I.1), about 163 km long, was inaugurated in 1869.

In contrast to the rift of the Gulf of Suez, the Gulf of Aqaba is wider, 19–27 km. The length of the Gulf of Aqaba as well as the length of the Asian border of the Peninsula with Israel between Rafah and the Gulf of Aqaba is about 200 km each.

Many different landscapes and habitats are found on the Peninsula. Topographically, the most conspicuous ones include:

- A complex mountain region around Gebel St. Catherine (2,641 m a.s.l.) in the south of the Peninsula, formed mainly by granite and diorite rocks. Typical for this region are high peaks [many of them reaching over 2,000 m a.s.l., such as G. Serbal (2,070 m a.s.l.), Al-Thabt (2,437 m a.s.l.), G. Umm Shomar (2,586 m a.s.l.) and G. Musa that is often identified as the biblical Mt. Sinai (2,300 m a.s.l.)] and deep large wadis (=gorges with fossil streams, which might be active intermittently) with steep slopes. Many wells and springs support small gardens in the area.
- A sandstone belt established northwards of the mountain region by large outcrops of hard Nubian sandstone. The sand derived from this sandstone fills wadis and covers the plains.
- The El-Tih (=maze) and Al-Ugma plateaus built up by nearly horizontal layers of chalks and marls broken with numerous outcroppings of limestone.

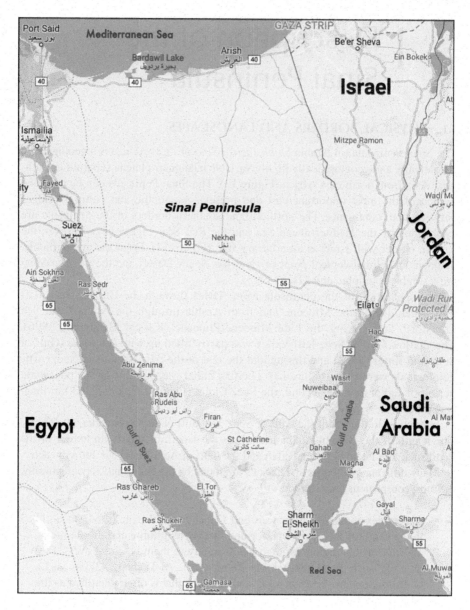

FIGURE I.1 The map of the Sinai Peninsula. (Downloaded from Google Maps: http://google.maps.com, and modified.)

These plateaus are drained mostly to the Mediterranean Sea by wide and shallow wadis.
- Coastal plains along the Gulf of Aqaba and the Gulf of Suez Rift Valley. The plain along the Gulf of Aqaba is narrower in comparison with the Suez coastal plain. The plain along the Gulf of Aqaba is covered by sediments

which originated from the magmatic and metamorphic rocks and, in part, from fossil coral reefs.
- Mediterranean coastal sandy plains (sabchas) including Lake Bardawil (a salty lagoon which is fed only by rainfall and by the Mediterranean Sea) (Zahran & Willis 1992). This rather flat, uniform landscape is characterised by quaternary deposits and sand dunes. Few anticlines bring landscape variability to this part such as the hulking Gebel Maghara (776 m a.s.l.), Gebel Halal (892 m a.s.l.), Gebel Libni (463 m a.s.l.) and Gebel Yiallag (1,094 m a.s.l.). Pastoralism and agriculture are mostly concentrated in the north-eastern region of the Peninsula. The most important drainage system is Wadi El-Arish which covers about 20,000 km^2 and extends over, approximately, 250 km (Zahran & Willis 1992).

1.2 CLIMATE AND PALEOENVIRONMENT

The present climate of the Sinai is arid and hyper-arid (Ayyad & Ghabbour 1986), characterised by large inter-annual and inter-local differences in the amount of precipitation.

Annual precipitation is up to 200 mm (Zahran & Willis 1992), and fog and dew significantly contribute to the water balance in some regions, especially in the high mountain ones. In addition, some mountain regions receive more than 300 mm of annual precipitation as snow (Zohary 1935). The water is stored underground in rock crevices or in the sand, and reaches the surface as springs and streams, or through artificial wells.

The minimum winter temperature ranges from 19°C in Sharm el Sheikh to 0°C in St. Catherine and the maximum summer temperature ranges from near 20°C in St. Catherine (altitude is about 1,586 m a.s.l.) to more than 50°C in El Kuntila (with an altitude of, approximately, 512 m a.s.l.) (Zahran & Willis 1992).

In winter, the potential evapotranspiration rate is not as high as in summer (in August, it exceeds 20 mm per day). Fortunately, most of the rains or snow in the mountains fall between November and April when temperature is relatively low, and evapotranspiration is not as high as in summer. Nevertheless, the water supply is very limited for both wildlife and people all year round.

The Sinai Peninsula has witnessed many climate changes in the past. The picture of these fluctuations is not clear yet because the cause-and-effect relationship between climatic changes and desertification, migration and invasion of species is not clear either. For example, a small lake existed in Gebel Maghara between ca. 12,500 and 11,500 BC (Goldberg 1977). Between ca. 5000 and 3000 BC, the locality, covered today by thorny shrubs, trees and only by a few grasses, was a region of steppes with savannah-like vegetation and settlements (Eddy & Wendorf 1999). Dendrochronological analyses of *Juniperus phoenicea* from G. Maghara and from G. Yiallag in the northern Sinai imply that the average amount of rainfall was 325 mm between AD 1540 and AD 1625, 39 mm in the period between AD 1625 and AD 1790 and this amount was about three times higher than in the previous dry period (Waisel & Liphschitz 1968).

Beetle reproduction phenology is driven largely by a temperature-photoperiodic controlled dormancy (Saunders 2021). The observation of pharate imagoes emerging from the soil in both autumn or spring gives us an indication, for example in darkling beetle *Pachyscelis rotundata* Kraatz, 1865 – T.P. personal observation – that it might be difficult for some beetle species to discern between the spring and autumn transitional seasons.

I.3 FLORA AND FAUNA

The combination of harsh climatic conditions, a long history of human settlements and the unique geographic position of the peninsula contribute to the current unique composition and stratification of flora and fauna in the Sinai Peninsula. This fauna and flora are not merely part of a more or less gliding transition from Asia to Africa and vice versa, but it has its own autochthonous endemism. Out of all taxa, most studies had been devoted to vascular plants, mammals and birds. Many invertebrate taxa have not been studied at all, for example, spiders.

I.3.1 Mosses and Vascular Plants

The flora of the peninsula comprises more than 1,285 species of higher plants (Anonymous 2002) and 27 rare moss species occurring at the mountain tops (Täckholm 1932). Montane alpine plants such as Sinai's primrose *Primula boveana* Decne, hawthorn *Crataegus sinaica* Boiss, *Pistacia khinjuk* Stock, Phoenicean juniper *Juniperus phoenicea* L. and many others seem to be relics from periods with different climate conditions in the Sinai. Among 30 of the mountain endemic species, *P. boveana* survives only near of permanent springs on the red granite at an altitude about 2,000 m a.s.l. This species originated from an ancestral species widely distributed in Africa and Asia during cool and wet period during late Miocene (about 6 MYA) (Wendelbo 1961). A remnant of the Irano-Turanian vegetation is *Pistacia khinjuk* Stocks (Danin 1983) growing in the cracks and on the cliffs in the southern mountains of the Sinai. It is also known from the Zagros Mountains in Iran and from Afghanistan.

The Mediterranean species *J. phoenicea* is abundant in the Mediterranean region while it is absent in Lebanon, Israel and Syria. It dominates in northern Sinai's regions getting more than 100 mm of rain per year such as Gebel Helal, Gebel El-Maghara and Gebel Yiallag. Contrary to *P. boveana* and *J. phoenicea*, the tree *Ziziphus spina-christi* represents a relic from previous African savannah-like vegetation periods. Today, this tree is limited to the more humid wadi bottoms where it is also grown by the local Bedouins who collect its edible fruits. Not unexpectedly, *Z. spina-christi* is mixed with other species of the African savannah origin such as *Vachellia tortilis* subsp. *raddiana (Savi) Kyal. & Boatwr. V. tortilis* subsp. *tortilis (Forssk.) Galasso & Banti* and other plant species in the alluvial fans at the foot of the mountains next to the southern coast of the peninsula. These fans contain abundant underground water.

Mangrove vegetation, characteristic of tropical and sub-tropical regions of the world, is to be found in the southern tip of the peninsula near of Ras Mohammad (near Sharm El Sheikh, Figure I.1).

Not only past but current climatic changes are also playing a role in the distribution of plant species in the Sinai. Geology and pedology are major factors too. Among others, the vegetation on the shore-line and around the salty lakes is composed of halophytes (salt-adapted species), for example: *Halocnemum strobilaceum* (Pall.) Bieb., *Arthrocnemum macrostachyum* (Moric.) C. Koch, common reed *Phragmites australis* (Cav.) Trin. ex Steud., sea rush *Juncus rigidus* Desf. and glycophytes (salt-tolerant species) like date palm, *Phoenix dactylifera* (Danin 1983). Mycorrhiza associated with them is helping them to tolerate the high concentration of salts in the soil. In the Sinai, the principal salts found in soils are gypsum (calcium sulfate dihydrate), calcium carbonate and sodium chloride (Danin 1983).

I.3.2 VERTEBRATES

The high diversity of plants is accompanied by a relatively high diversity in some animal taxa.

The mammalian wild fauna is characterised by not less than 12 species, among others, rocky hyrax, *Procavia cappensis* (Pallas, 1766), ibex, *Capra nubiana* Cuvier, 1825, leopard, *Panthera pardus* (Linnaeus, 1758) and striped hyena, *Hyaena hyaena* (Linnaeus, 1758) (Anonymous 2002). Not to be overlooked are the feral mammals that might have negative effects on the biodiversity of the Sinai Peninsula, such as feral donkeys *Equus africanus asinus* Linnaeus, 1758 and dromedaries *Camelus dromedarius* Linnaeus, 1758.

The bird fauna of the Sinai seems to be underestimated because only 45 bird species (sedentary and migratory ones) had been recorded in the southern Sinai (Baha el Din 1999). There were, among others, white and black storks, *Ciconia ciconia* (Linnaeus, 1758) and *Ciconia nigra* (Linnaeus, 1758), Tristram's grackle, *Onychognathus tristrami* (Scatler, 1858), sparrow hawk, *Accipiter nisus* (Linnaeus, 1758), common buzzard, *Buteo buteo* (Linnaeus, 1758), long-leg buzzard, *B. rufinus* (Cretzschmar, 1829), golden eagle, *Aquila chrysaetos* (Linnaeus, 1758), the western Palearctic Verreaux's eagle, *A. verreauxii* (Lesson, 1831) and kestrel, *Falco tinnunculus* (Linnaeus, 1758) (Baha el Din 1999).

In the Sinai, we find 23 lizard species (Anonymous 2002), more than 20 species of snakes (Anonymous 2002) and 5 amphibian species. The amphibians are: Reuss's toad, *Sclerophrys regularis* (Reuss, 1833), thirsty frog (toad), *Bufotes sitibundus* (Pallas, 1771) (mentioned as *B. variabilis*), *Hyla savignyi* Audouin, 1827, Bedriaga's frog, *Pelophylax bedriagae* (Camerano, 1882) and Nile grass frog, *Ptychadena nilotica* (Seetzen, 1855) (Borkin et al. 2016). It is expected that all the amphibian species were introduced in the past 20–30 years in connection with land cultivation and irrigation (Borkin et al. 2016).

I.3.3 INVERTEBRATES

As for the invertebrates of the Sinai, only a few taxa, such as butterflies (Larsen 1990), earthworms (Pavlíček et al. 1997) and darkling beetles (Lillig & Pavlíček 2003), have up to now been the subject of faunistic publications.

Out of 44 known butterfly species recorded in the Sinai, 34 reproduce on the peninsula and the rest are migratory butterflies and/or have an unknown breeding status (Anonymous 2002). Out of these 44 species, two are endemic species, Sinai hairstreak *Satyrium jebelia* (Nakamura, 1975) and Sinai baton blue *Pseudophylotes sinaicus*

Nakamura, 1975. *P. sinaicus* is limited onto an area above 1,500 m a.s.l. where it feeds exclusively on Sinai thyme *Thymus decussatus* (Bentham) (Anonymous 2002).

Regarding darkling beetles, see Lillig & Pavlíček (2003) and this book.

I.4 LOCAL GEOGRAPHIC NAMES

There are usually no linguistic problems with the reference to the Sinai Peninsula, during the period of beetle collections. This territory has been referred to mostly as "Sinai" or "Sinai Peninsula" during both the late Ottoman period and after the peninsula had been attached to Egypt in 1922. However, relative many specimens from the Sinai had been labelled "Israel". This refers to the period of the Israeli military control of the Sinai Peninsula during the period 1967–1982 following the Six-Day War of 1967, and perhaps also during 1956–1957 (Uhmann et al. 2005) following the Tripartite aggression of 1956.

The location of the place names (toponyms) of the collection sites mentioned in the literature or on the labels often proved to be difficult. In general, this is reflected in the fact that 22.4% of the toponyms in the Sinai have unknown meaning (Zeini et al. 2018). The mono- and multi-meaning names contribute by 64.1% and 13.5%, respectively. The difficulties in the localisation of the toponyms are due to the usage of different transliterations of geographic names from Arabic into different languages. In our case, apart from Latin, we deal with transliterations into English, French, German and Italian. For example, the transliteration of the Arabic word *waha* or واحة into English is *oasis*, into French it is *oasis*, into German it is *Oase* and in Italian it is *oasi*. Furthermore, the mixing of a few dialects of spoken Arabic and historic-induced changes into the local names, as for example Gabal (=mountain) in the Arabic Cairene dialect may be pronounced and written Jabal or Jebel in Bedouin dialects. In some cases, the current toponyms originated from biblical sources, or from the period of the Roman and Greek settlements and were modified in the Arabic languages such as, among others, the change of the name of a Greek settlement name Pelusium into Romani (or Rommanach and Rodmano). We also discovered a few potential cases of geographic homonyms (two or more toponyms that have identical spelling and pronunciation). For example, the geographic name Wadi Musa (valley of Moses, sometimes written as Mussa), its spring Ain Moses (Moses's water spring or well) and Aaron tomb are localised in the Sinai as well as in the Wadi Musa in southern Jordan.

It might be difficult to distinguish whether changes in the name writing represent a homonym, an error in the toponym translation or simple name alternatives. As a matter of fact, without knowing the geographic character of the named place, we cannot investigate the motive behind naming it. In order to not increase the chaos in the Sinai toponymy, we left the names in their original writing, as published or written on labels.

Here, we list the most frequent synonyms and variants of the toponyms in the Sinai Peninsula (see Geobase.com for more details): Abas Bosha (=Qasr Abbas Pasha); Abu Rudeis (=Abu Rhodes, =Abu Rueitat); Abu Zeneima (=Abou Zenneh, =Abu Zenima); Ain Furtaga (=Eim Furghaga); Ain Hawara (=Hawara); Ain Hudra (=A. el-Houdra, =A. al-Houdra, =A. Khudra); Aoul Haibeh (=Aoual Heneibeh, =

Description of the Sinai Peninsula

Auel Habesh); Ayun Mussa (=Aium Mussa, =Ain Musa, =Oase Ain Musa); Bir Abu Rudais (=Bir Abu Rudeis); Bir Tmade (=Bir el Thamada); Dhahab (=Dahab); El Quseima (=Kuseima); Gebel El Tih (=Djebel El Tih, =Jebel El Tih, =Djebel el Hih, =Djebel Hamr); Gebel Hallal (=G. El Helal); Gebel Hammam Firaun (=Hamman Fir'aoun, =Djebel Hamam, =Gebel Hammam Faraun, =Pharaoh's Baths); Gebel Katherin (=Mt. Catherine); Gebel M'rara (=G. Maghara in the Northern Sinai); Gebel Maghara (=Meghara east of Abu Rudeis, not the one in the northern Sinai); Gebel Mussa (=Mt. Moses, Mt. Sinai, Berg Sinai); Gebel Sefrou (=Djebel Sefrou); Gebel Serbal (=Djebel Serbal); Gebel Watiyeh (=Djebel Ouatiyet, =D. Watiyeh); Hadjar er-Reqab (=Hajar er-Rekkab); Monastery of Santa Katharina (=Katharinenkloster, St. Katharina); Kuntilla (=El Kontella, =El Kuntilla); Ma'agana (=Ma'agama); Mazar (=El Mazar); Mitla Pass (=Mitla, =Mitle); Mohammedia (=Mohamdiya, = Gerrha); Mount (=Gebel, =Jebel, =Djebel); Nakhl (=Nachle, =Qala'at en-Nakhel); Naqb el-Haoua (=N. el Haoula, =N. el-Hasua, =Nakb Hawa), Nuweiba (=Nouheibeh); Oase el-Hessoueh (=O. el Hosseyeh); Om Bogma (=Umm Bugma, =Bugnia); Pharan (=Oase Feiran, =Refidemn, =Refidim); Qantara (=El Qantara); Rafa (=Rafiah); Reidan Esqua'ah (=Ridhan Schkaa); Scheik Hamid (=Cheik Hamid); Scheik Nabi Salah (=Scheich Nebi Saleh); Serabit El Khadem (=Serabit El Hadam); Sobchat A-Sheck (=Sabkhet el Sheikh, =Sebchat A-Sheck); Wadi Ain El Gederat (=W. el Gedeirat); Wadi Amara (=W. el-Amara, =W. Amarah); Wadi Baba (=W. Ba'baa); Wadi Budr (=W. Budra); Wadi El Scheikh (=W. Cheikh, =W. es Scheich); Wadi El Tarfa (W. Tarfa); Wadi eth-Thal (=W. Thal, =W. Attal (Athal)); Wadi Genneh (=W. Gneh); Wadi Gergir (=W. Girgir); Wadi Gharandel (=W. Gorondel); Wadi Khamila (=W. Khamileh, =W. Chamile); Wadi Kid (=Kid); Wadi Mokatteb (=Mokatteb); Wadi Qnai el Rayan (=W. Kunel-el-E-Rayah, =W. Qnai el-Royan), Wadi Rachaba (=W. Rahabeh, =W. Rahaba), Wadi Rayane (=W. Riene, =W. Rieina); Wadi Sa'al (=W. Sa'l); Wadi Schellal (=Shellal); Wadi Sidri (=W. Sidr); Wadi Solaf (=W. Selaf); Wadi Taba (=W. Tabah); Wadi Tayebeh (=W. Taiyibeh); Wadi Tlach (=W. Tlah); Wadi Um Mitla (=W. Mitla); Wadi Wardan (=W. Werdan, =W. Ouerdan); Plain of Ga'a (=Ga'a Wüstenebene, =El Ga'a, =Ga'a, =Wüste Ga'a, =Ebene Ga'a).

II Collection Methods and Analysis of the Darkling Beetles Distribution

The collection sites of darkling beetles are unevenly distributed over the Sinai Peninsula and are concentrated along the roads crossing the peninsula. Collectors have most frequently used the road from Suez to Abu Zeneima or Abu Rudeis and from there passing Serabit el Khadem and Wadi Feiran to the central mountain massif around the Monastery of St. Catherine or the road which leads along the Gulf of Aqaba from Taba (on the border with Israel) to Dahab, and from there to St. Catherine. Less frequently, collections have been done along the "Hadj" road from Suez to Taba via Nakhl, along the "camel road" from Tor to St. Catherine or along the Sinai part of the Via Maris (Philistine Road) along the Mediterranean coast.

Tenebrionid beetles have been collected by watching the surface for movements of their adults, watching under stones, logs and bark of dead or dying wood, examining other objects covering the soil or sand surface, and by digging into the upper layer of the sand. Checked, for occurrence of darkling beetles, were also fish remains, and piles of algae accumulating on the beach due to tidal sea activity, and by using pitfall traps (Norfolk et al. 2012).

We only conducted the preliminary analysis of the differences in the distribution of each species of the darkling beetles between the North and South Sinai. In the absence of available abundance data, we decided to use a binomial test in order to test the difference in the number of reported collection sites between South and North Sinai. Therefore, the working assumption was that the difference in the number of collection sites is positively paralleled by the prevalence in each species. We chose to test this assumption by using a binomial test (https://www.socscistatistics.com/tests/binomial/default2.aspx).

In order to compensate for the different number of trials (n) between the north and south parts of the Peninsula, we calculated the statistical parameter p (the proportion between the total number of collection localities as reported in the South and North Sinai). Therefore, $p = 0.33$ is the probability that the corresponding outcome will appear in every occasion (=at every species or subspecies) and $q = (1 - p) = 0.67$. Both parameters p and q indicate that two-thirds of the localities are present in South Sinai and one-third occurs in North Sinai. The other value "K" determines the number of trials with the positive outcome of finding the beetle species at the North or South Sinai. Table II.1 presents a significant outcome in ten species with a higher number of collection localities in the South Sinai.

TABLE II.1
Darkling Beetle Species That Significantly (Binomial Test, $p > 0.05$) Differ in Their Prevalence between South and North Sinai: The Species Are Arranged in Alphabetical Order

Species Name / Map Number	Number of the Recorded Collection Sites in South and North Sinai	$n/p/q/K/P$
Adesmia cancellata abbreviata / Map VI.2	40 / 7	47 / 0.33 / 0.77 / 7 / 0.003
Adesmia cothurnata cothurnata / Map VI.3	35 / 3	39 / 0.33 / 0.77 / 3 / 2.10^{-4}
Adesmia montana montana / Map VI.1	8 / 0	8 / 0.33 / 0.77 / 0 / 0.041
Erodius puncticollis sinaiticus / Map VI.4	16 / 2	16 / 0.33 / 0.77 / 2 / 0.03
Mesostena angustata angustata / Map VI.5	18 / 1	18 / 0.33 / 0.77 / 1 / 5.10^{-4}
Opatroides punctulatus punctulatus / Map VI.7	18 / 2	22 / 0.33 / 0.77 / 2 / 0.02
Oxycara peyerimhoffi / Map VI.6	15 / 1	18 / 0.33 / 0.77 / 1 / 0.01

III Species Richness and Endemism of Tenebrionidae Beetles in the Sinai Peninsula

Based on the data presented here, we regard 127 species of darkling beetles as an integral part of the fauna in the Sinai Peninsula. *Nota bene*, we consider that *Adesmia metallica* has two subspecies in the Sinai, so that the total number of valid species and subspecies recorded in the Sinai is effectively 128. However, the distribution and taxonomic positions of *Zophosis bicarinata* cf. *ghilianii*, *Drosochrus costatus aegyptiacus*, *Cheirodes sardous sardous*, *Phtora apicilaevis* and *Ph. subclavata* are taxonomically unknown, unclear or uncertain.

An additional 17 species might, in our opinion, live at the peninsula, but their presence needs to be confirmed. It means that the expected number of the darkling beetles for the Sinai is in the range between 122 and 144 species. We discuss here also 43 taxa (species and subspecies) that are mentioned in the literature, but should, according to us, not be regarded as part of the Sinai fauna as long as their presence in the region is not verified.

For comparison:

Malta	316 km^2	63 taxa (Grimm et al. 2020; Lillig et al. 2012)
Qatar	11,500 km^2	51 taxa (Soldati 2009)
Sicily	25,000 km^2	ca. 160 taxa (Aliquò & Soldati 2010)
Sinai	61,000 km^2	128 taxa

III.1 GLOBAL DISTRIBUTION PATTERNS OF THE TENEBRIONIDAE IN THE SINAI PENINSULA (FIGURE III.1)

The classification into "global chorotypes" represents the world-wide spatial responses of species to historical and environmental pressures, and does not vary with the area under consideration (Fattorini 2015). Around 72% (13 out of 18) of the used categories are based on Vigna Taglianti et al. (1999). These are identified by the numerical codes (in brackets) used in this chapter. As already noted by Soldati (2009), this categorisation is incomplete and numerous species cannot be classified

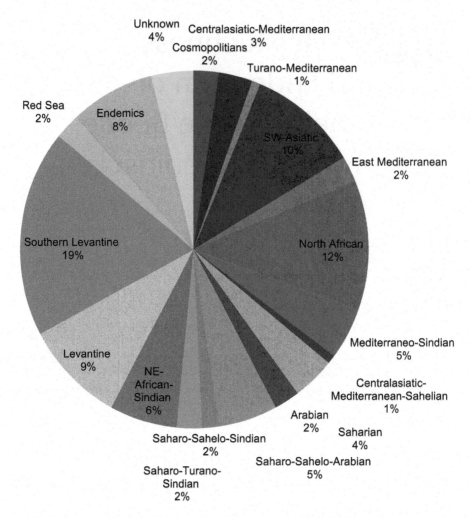

FIGURE III.1 Representation of different chorotypes of darkling beetles on the Sinai Peninsula delineated according to their general distribution.

in this system. For these, some of the global chorotype categories mentioned by Lillig and Pavlíček (2003) are retained.

- **Cosmopolitan taxa (0.01)**
  (*Alphitobius diaperinus, Tribolium castaneum, Tribolium confusum*)
  Cosmopolitan species occur in all zoological regions.
  So far, only three cosmopolitan species are known from the Sinai. They are displaced by humans and usually treated as pests. Their region of origin is usually difficult to locate. *Alphitobius diaperinus* is a tropical fungivorous

species. In the temperate zone, the species is regularly found in pigsties and hen-houses as well as in kitchens and bakeries (Pospischil 1996). *Tribolium confusum* originates from Africa, and *T. castaneum* is an Indian species (Hinton 1948). Both are known as pests of stored products (Hinton 1948).

- **Centralasiatic-Mediterranean chorotype (1.08)**
  (*Zophosis punctata punctata, Gonocephalum rusticum, Scleropatroides hirtulus, Cheirodes brevicollis*).
  Widespread species from the Gobi Desert to the Aralo-Caspian depression, Middle East, Caucasus, Anatolia, North Africa, southern Europe.

- **Turano-Mediterranean chorotype (1.11)**
  (*Ammobius rufus*)
  Widespread in southern Europe, North Africa, Middle East, Iran, Anatolia, Caucasus, west Turkestan.

- **SW-Asiatic chorotype (1.13)**
  (*Adesmia montana montana, Amnodeis confluens confluens, Pimelia hirtella, Mitotagenia aegyptiaca, Dailognatha crenata, Micipsa douei douei, Micipsa philistina, Zophosis complanata, Zophosis pharaonis pharaonis, Blaps judaeorum, Blaps wiedemanni, Proscheimus arabicus, Omophlus gracilipes*).
  Widespread species in Middle East countries, eastern Mediterranean coastal regions from Anatolia to Sinai, from Arabian Peninsula to Iran and Caucasus.

- **East Mediterranean chorotype (3.03)**
  (*Sepidium dathan, Hionthis tentyrioides, Tentyrina orbiculata subsulcata*)
  Species of the eastern countries of the Mediterranean basin from Italy and Sirte (Libya) to the Black Sea with extensions to the Middle East and to the Sahara Desert.

- **North African chorotype (3.04)**
  (*Adelostoma cordatum, Machlopsis crenatocostata, Erodius zophosoides zophosoides, Pimelia barthelemyi, Pimelia canescens canescens, Trachyderma genei, Sepidium tricuspidatum tricuspidatum, Oxycara pygmaeum, Tentyria punctatostriata, Zophosis nigroaenea, Zophosis lethierryi, Blaps bifurcata bifurcata, Blaps nitens laportei, Blaps polychresta*).
  Species distributed in North Africa north of the Sahara with extensions to neighbouring areas. Many North African species are distributed towards the east and are reaching the Sinai or the Negev Deserts. Some of these species are reaching the Cyrenaica in the west, also the species that appear westwards only up to Libya.

- **Mediterraneo-Sindian chorotype (3.05)**
  (*Adelostoma sulcatum sulcatum, Trachyderma philistinum, Gonocephalum perplexum, Opatroides punctulatus punctulatus, Leichenum pulchellum pumilum, Scaurus aegyptiacus*).

    Taxa of this category are widespread in the Mediterranean countries. To the east, their range extends to the Sind (Pakistan). In the case of *O. punctulatus punctulatus*, the range extends to Kazakhstan and Kashmir.

- **Centralasiatic-Mediterranean-Sahelian chorotype**
  (*Gonocephalum setulosum setulosum*)

    The species of this chorotype are widespread over large parts of northern Africa to central Asia as the Centralasiatic-Mediterranean chorotype, but extends southwards into the Sahel.

- **Saharian chorotype (5.05)**
  (*Pterolasia squalida, Scelosodis castaneus castaneus, Zophosis personata, Zophosis plana, Mycetocharina megalops megalops*).

    Species widespread in the Sahara from the Atlantic Ocean to the Red Sea. *Pterolasia squalida* also occurs disjunctively distributed in Senegal.

- **Arabian chorotype (5.06)**
  (*Akis spinosa, Pimelia bottae, Vieta tuberculata*)

    Species occurring in the Arabian Peninsula from the Gulf of Aqaba to the Persian Gulf.

    The records of *Pimelia bottae* in the Sinai lie somewhat outside the Arabian chorotype. Nevertheless, the species is assigned to this type.

- **Saharo-Sahelo-Arabian chorotype (5.08)**
  (*Prionotheca coronata coronata, Trachyderma hispidum, Cyphostethe heydeni, Mesostena angustata angustata, Trichosphaena perraudierei perraudierei, Sclerum orientale orientale, Cheirodes pilosus*).

    Widespread over the Sahara Desert, the Sahel and the Arabian Peninsula.

- **Saharo-Sahelo-Sindian chorotype (5.09)**
  (*Gonocephalum soricinum soricinum, Cheirodes asperulus*)

    As the Saharo-Sahelo-Arabian chorotype, but extending further eastwards to the Sind.

- **Saharo-Turano-Sindian chorotype (5.10)**
  (*Imatismus villosus, Clitobius oblongiusculus, Gonocephalum patruele patruele*).

    This chorotype includes species spread from the Sahara to the Aralo-Caspian Depression and eastwards to the Sind.

- **NE-African-Sindian chorotype (5.11)**
  (*Adesmia cothurnata cothurnata, Thriptera crinita crinita, Thriptera kraatzi, Mesostena puncticollis, Prochoma audouini, Tentyrina palmeri palmeri, Zophosis farinosa, Phaleria prolixa*).

The species occur in north-east Africa from Egypt east of the Nile to Somalia, in Asia in Arabia and in southern Iran to the Sind.

- **Levantine chorotype**
  (*Akis reflexa, Erodius dejeanii, Erodius duponchelii, Erodius gibbus gibbus, Pimelia angulata angulata, Pimelia mittrei, Pimelia subquadrata subquadrata, Micipsa burtoni, Tentyria discicollis, Dendarus syriacus, Mesomorphus setosus, Scaurus rugicollis*).
  The species distributed from Israel or Lebanon, sometimes Syria, to Egypt, belong to this chorotype.

- **Southern Levantine chorotype**
  (*Adesmia metallica brozai, Adesmia metallica laevior, Adesmia dilatata dilatata, Erodius kneuckeri kneuckeri, Erodius opacus, Erodius hebraicus, Erodius puncticollis sinaiticus, Astorthocnemis becvarorum, Pimelia arabica edomita, Pimelia theveneti, Thriptera varvasi pilipes, Microtelus careniceps binodiceps, Stenosis affinis, Hegeterocara arabica, Micipsa schaumii, Oterophloeus alveatus peyerimhoffi, Oterophloeus haagii orientalis, Oxycara ardoini, Oxycara peyerimhoffi, Thraustocolus leptoderus, Eurycaulus henoni, Hedyphanes chikatunovi, Pseudoseriscius maculosus murinus, Cteniopus priesneri luteiceps*).
  These species are distributed on a much smaller area than the Levantine species: from Lower Egypt to southern Jordan. Many species are living only in the Sinai and the Negev deserts. We think that these species should not be called "Levantine" since they are present only in the southern Levantine desert regions, whereas "true" Levantine species are present in the desert as well as in mesic regions.

- **Red Sea seashore chorotype**
  (*Clitobius grimmi, Diphyrrhynchus aenescens, Trachyscelis tenuestriatus*).
  These three species are known only along the Red Sea coast.

- **Sinai endemics chorotype**
  (*Adesmia cancellata abbreviata, Trachyderma andresi, Dichillus alfierii, Mesostena peyerimhoffi, Mesostena picea sinaitica, Oxycara productum, Schweinfurthia alfierii, Schweinfurthia sinaitica, Tentyria sinaitica, Dendarus calcaroides*).
  The Sinai endemics category includes the species that were found only in the Sinai Peninsula and not anywhere else.

- **Unknown**
  Distribution and taxonomic status of **Z. bicarinata cf. ghilianii** are unknown. **Drosochrus costatus aegyptiacus**: The distribution data in Iwan et al. (2020) appear uncertain. The status of **Cheirodes sardous sardous, Phtora apiciaelvis** and **Ph. subclavata** are unclear.

The Tenebrionidae of the Sinai Peninsula are composed of species with numerous distribution patterns. Striking is the high proportion of species that have a relatively small distribution area. Thus, well over one-third of all species recorded on the Peninsula are among the endemics (8%) or those with Levantine distribution (10%) and Southern Levantine distribution (19%). In contrast, the extensively spread species, such as cosmopolitan species, are much less represented. However, more species of the latter group are very likely to be discovered in the future, especially in settlements.

In our opinion, our data indicate that the Sinai Peninsula served for darkling beetles as a continental bridge between northern Africa and the Levant and Arabia. In fact, we were surprised to find a substantial number of African species in the Sinai Peninsula. At present, we do not know whether this pattern reflects the situation that existed before the separation of the Sinai Peninsula from Arabia (Late Miocene/ Pliocene: Kuss & Bachmann 1996) or whether the exchange between the Sinai and Arabia took place due to an active (flight) or passive (e.g. wind drift) migration over the Gulf of Aqaba or due to an active migration along the coast of the Gulf of Aqaba. Nevertheless, the Sinai Peninsula is also a "crossroad" where many North African species have their eastern border of distribution and many Asian, Levantine and northern Red Sea species their western one. Interestingly, the number of species coming from Asia to the Sinai is not much lower than the portion of taxa reaching the Peninsula from Africa. This indicates that Africa and Asia contribute to the darkling beetle fauna of Sinai almost in an equal proportion. It should also be noted that no tenebrionid species of Ethiopian and/or tropical origin are present in the Sinai, contrary for example to Trichoptera (Botosaneanu 1999) and butterflies (Larsen 1990).

Last but not least, the Peninsula seems to be a place of autochthonous speciation, as indicated by the presence of endemic species. Furthermore, species present only in the deserts of the southern Levant and along the coast of the Gulf of Aqaba originated perhaps from the Sinai Peninsula. The fact that about one-third of the species present on the Sinai Peninsula has small distribution areas, calls for the preservation of their habitats. The restriction of most endemic species of the Sinai to the mountainous area in the southern part of the Peninsula should also be taken into consideration with regard to their protection. As judged from the small portion of cosmopolitan species (3%), the Levantine fauna is not yet much influenced by recently introduced species.

IV Ecology of Desert Darkling Beetles

The Sinai Peninsula has a very small-scale landscape structure. It provides habitats for a whole range of ecologically differently adapted species. Halophilous species such as *Ammobius rufus*, *Phaleria prolixa* and *Diphyrrhynchus aenescens* inhabit the coastal regions. The sandy areas in the north and some wadis are inhabited by sabulicolous species such as *Erodius* spp. Typical species of the rugged mountains are *Dichillus alfierii* and *Tentyria sinaitica*.

The extreme climatic conditions of the desert with high solar radiation, a large day-night temperature amplitude, severe aridity as well as variability in the edaphic conditions led to numerous adaptations of the animals living there. Special adjustments are also found in the Tenebrionidae, the beetle family that dominates in all sub-tropical deserts.

A striking feature of desert tenebrionids, including those of the Sinai Peninsula, is the high proportion of nocturnal species. The vast majority of the Tenebrionidae of Sinai are nocturnal and spend the day buried in the sand or under stones to escape the sunlight and high temperatures. Nevertheless, specimens of *Adesmia*, *Erodius* and *Pimelia* spp. can also be observed active during the day. Some species are protected by a light blue wax layer. It is conspicuous in the genera *Erodius* and *Zophosis*.

Of particular importance for all animals is the lack of water. This is also true for the Tenebrionidae. A widespread strategy is flightlessness. In many species, for example in the Pimeliini, the elytra are firmly fused together, which makes it difficult for the abdomen to dry out. Exceptions to flightlessness are for instance the genera *Cyphostethe*, *Imatismus* or *Cheirodes*. These species are nocturnal and are occasionally attracted to light at night. The authors observed this at the Jordan–Israel border at the end of April 2004, where a large number of *Cheirodes* specimens flew past the lights at the customs post.

In order to reduce water loss from the body, tracheae in many species, e.g. *Blaps*, open into a sub-elytral cavity. The function of this chamber as a protection against evaporation is undisputed. The cavity significantly reduces the evaporation rate. Originally, it was also assumed that it also serves as a protection from solar radiation. However, Cloudsley-Thompson (1965, 1978) could not find any differences in temperature outside and inside the sub-elytral cavity in experiments with *Pimelia grandis grandis* from the Sudan.

Some species are known to be able to replace the partial loss of an elytra with secondary sclerotisation if they survive the first month after injury (Crowson 1981).

The larvae of many desert tenebrionids are protected from dehydration by a hard sclerotised cuticula (Crowson 1981).

The adult Tenebrionidae also developed morphological adaptations to edaphic conditions. Some species (e.g. *Pimelia angustata angustata*, *Astorthocnemis becvarorum*) have a long pubescence on the tarsi that increases their surface area. In this way, they act as "snow shoes for the sand" and prevent the beetles from sinking between the grains of sand. Pierre (1958) establishes a connection between the length of the pubescence and the structure of the sand.

Diurnal species, such as some *Adesmia* spp., face the problem of the heat radiation coming from the substrate. The long legs are interpreted as "spacers". The further the body is from the ground, the less it is affected by the heat radiated from it. Like, on a different size scale, in Arabian camels (Hoter et al. 2019).

Many burrowing species have at least converted the front legs into burrowing legs. Representatives of the genus *Erodius*, for example, are able to burrow into the loose sand, usually near plants, at lightning speed.

The species of the genus *Zophosis* also dig quickly, although they have no burrowing legs. The dytiscidae-like shape of their bodies helps them to do this. *Zophosis* species practically dive into the sand. When diving, they move their legs synchronously as if they were swimming breaststroke. This way, they are faster than with alternating movements like those of a crawl swimmer (Crowson 1981).

Another problem for desert-dwelling animals is obtaining food, which is usually scarce. Tenebrionidae are not choosy. They are considered omnivores. Mostly they feed on blown, dry plant material. They also do not disdain carrion and excrements. With this kind of food, they consume enough water and do not need any other water sources. However, if possible, they also climb on plants and enjoy fresh green leaves or tasty flowers (Grimm 1983).

Some tenebrionid species are considered myrmecophilous, and such species can be found in the genera *Dichillus, Eutagenia, Palorus, Myrmechixenus* (Lapeva-Gjonova 2013) or *Mesostena* (Soldati 2009). *Dichillus* and *Mesostena* are also found in the Sinai.

Specimens of *Mesostena angustata* were observed grabbing dead insects and other food transported to the nest by ants of the species *Messor intermedius* Santschi, 1927. When the food supply decreased, *M. angustata* individuals suddenly attacked *M. intermedius*. A group of *M. intermedius* specimens fought back and attacked the *M. angustata* specimens, but the beetles' hard chitinous integument protected them from the ants' attacks (Makhan et al. 2011).

V Identification Keys for the Tenebrionidae Taxa Recorded on the Sinai Peninsula

KEY 1 IDENTIFICATION KEY TO THE SUBFAMILY LEVEL

1 Claws never pectinate 2

1' Claws of tarsi pectinate (Alleculinae) .... Key 5

2 Base of antennae not covered by the lobe-like extension of the genae, the insertion site of the antennae visible from above .... Lagriinae
(only *Lagria* sp. is published from the Sinai)

2' Base of antennae covered by the lobe-like extension of the genae, antennae arising from beneath the genae, the insertion site of the antennae not visible from above 3

3 Abdomen without visible membrane at posterior margins of ventrites three and four 4

3' Abdomen with sometimes difficult to detect membrane at posterior margins of ventrites three and four 5

4 Anterior edge of epistome deeply emarginated; protibia widened apically .... *Cheirodes* (see Key 3)

4' The anterior edge of epistome is not emarginated but rounded or truncate or dentate (Pimeliinae pars) .... Key 2

5 Protibia distally widened, outer angles rounded; antennae sometimes gradually broadened apically (Diaperinae) .... Key 4

5' If protibia widened distally, the outer corners are angled; antennae never gradually broadened apically 6

6 Size more than 10 mm; with pilosity at least partially; antennae bristled; terminal joint of antennae small; terminal joint of maxillary palps not axe-shaped; base of pronotum almost always margined; anterior tarsi of ♂ not broadened (Pimeliinae pars: Pimeliini) .... Key 2

6' Characters not combined (Blaptinae and Tenebrioninae) .... Key 3

KEY 2 IDENTIFICATION KEY FOR THE SUBFAMILY PIMELIINAE

1. No membrane between third and fourth ventrite visible 20
1'. Membrane between third and fourth ventrite clearly visible (Pimeliini) 2
2. Elytra double-hairy, dense, carpet-like, grey appressed and with long red, erect hairs, only the low costae translucent black; anterior edge of protibia sharp; prosternal apophysis projecting horizontally to behind the procoxae .... *Pterolasia squalida*
2'. Characters not combined 3
3. Metatibia round or oval 12
3'. Metatibia sharply carinate and posterior face flattened, prosternal apophyses bent (*Pimelia*) 4
4. Metatibia with dense, light-coloured hairs on the flattened posterior side 5
4'. Metatibia glabrous 7
5. Pronotum on disc smooth or finely punctate, without granules .... *Pimelia canescens canescens*
5'. Pronotum granulated at least along the anterior or posterior edges 6
6. Pronotum almost 3 times as wide as long, strongly granulated at the middle .... *Pimelia mittrei*
6'. Pronotum almost 2.5 times as wide as long, sparsely granular on middle, almost smooth except for a few granules; interstitiae convex .... *Pimelia bottae*
7. The inner dorsal row of the elytra basally extinguished or smoothed 8
7'. The inner dorsal row of the elytra extends as single, widely spaced grains up to base 9
8. Elytra long protruding hairy; elytral epipleura fine and densely granulated .... *Pimelia arabica edomita*
8'. Elytra not long protruding hairy; elytral epipleura glabrous .... *Pimelia angulata angulata*
9. All primary rows of the elytra, including the lateral one, of sharp, pointed, conical and vertically projecting teeth .... *Pimelia theveneti*
9'. Primary rows formed posteriorly from small, inconspicuous tooth-like granules 10

10   Metatarsus long hairy, on the middle tarsomeres twice as long as these, on the basal tarsomer as long as this; elytra long oval, flattened above, on the basal half long projecting and densely hairy .... *Pimelia subquadrata subquadrata*

10' Pubescence of the metatarsi shorter, on the middle tarsomeres at most a little longer than these; the two inner rows of the elytra also behind hardly emerging from the surrounding elytra sculpture 11

11   Elytra strongly rounded, widest behind the middle; lateral row up to the base strong and densely granulated .... *Pimelia hirtella*

11' Elytra less broad, widest behind the middle; lateral primary row at base very sparse and finely granular .... *Pimelia barthelemyi*

12   Integument covered with round scales; elytra with two costae each; anterior tibia with 4–6 mm long spines; anterior margin of mentum excavated, bordered above and below by one ridge each .... *Astorthocnemis becvarorum*

12' Integument not covered with scales; protibia without any large spine, mentum not excavated 13

13   Elytra with a ring of spines laterally; median and metatibia with spines; median and metatarsi with long hairs .... *Prionotheca coronata coronata*

13' Elytra and tibia without any spines 14

14   Lateral margin of the head does not form an indentation to the anterior margin of the epistome (*Thriptera*) 15

14' Lateral margin of the head forms an indentation towards the anterior margin of the epistome (*Trachyderma*) 17

15   Beside the protruding hairs of the elytra with woolly recumbent hairs .... *Thriptera varvasi pilipes*

15' Elytra without recumbent pubescence 16

16   Prosternal apophysis far exceeds the anterior coxa to below the mesosternum; aedeagus compressed laterally in its apical part .... *Thriptera kraatzi*

16' Prosternal apophysis distinctly shorter, not reaching the mesosternum; aedeagus not compressed laterally in its apical part .... *Thriptera crinita crinita*

17   The inner margin of the elytral epipleura is slit-shaped anteriorly; prosternal apophysis always bent down behind the coxa .... *Trachyderma hispidum*

17' The inner margin of the elytral epipleura not slit-shaped; prosternal apophysis protruding the anterior coxa a bit 18

18 Pronotum double haired, covered with grey-white toment between the protruding hairs .... *Trachyderma philistinum*

18' Pronotum without a toment 19

19 Prosternal apophysis, seen from the face, with converging apex, but completely rounded at the apex; seen in profile it is slightly curved towards the abdomen, with subparallel sides and blunt vertex, almost completely rounded; elytra slightly wider than pronotum .... *Trachyderma andresi*

19' Prosternal apophysis, examined from the face, with very acute apex, examined in profile it is conical, straight or curved with very acute apex; elytra much wider than pronotum .... *Trachyderma genei*

20 Mentum large, without a cleft between mentum and buccal fissure lobe 21

20' Mentum smaller, with a cleft between mentum and buccal fissure lobe 77

21 Metacoxa not oblique; metasternum without longitudinal sulcus 30

21' Metacoxa strongly oblique; metasternum with an incomplete longitudinal sulcus (Zophosini) 22

22 Elytra without costa 24

22' Elytra with costa 23

23 Dorsal costa much closer to suture than to humeral costa .... *Zophosis plana*

23' Dorsal costa at most slightly closer to suture than to humeral costa .... *Zophosis bicarinata* cf. *ghiliani*

24 Pseudopleural crest distinctly sinuate posteriorly 25

24' Pseudopleural crest distinctly not or barely sinuate posteriorly 26

25 Granules on pseudopleura rounded to briefly elongate, strioliform granules rare .... *Zophosis complanata*

25' Pseudopleura with strioliform granules .... *Zophosis farinosa*

26 Elytra elongated, evenly transversely curved, without longitudinal bulges along the suture, suture not impressed 27

26' Elytra scarcely longer than wide, with longitudinal bulges along the impressed suture .... *Zophosis pharaonis pharaonis*

Identification Keys for the Tenebrionidae Taxa 23

27 In apical view pseudopleural crest completely visible in apical third 28

27' In apical view pseudopleural crest not completely visible in apical third, as it is arched over by the elytra 29

28 Genae in dorsal view broadly rounded; suture of elytra slightly impressed .... *Zophosis personata*

28' Genae in dorsal angular; suture of elytra not impressed .... *Zophosis lethierry*

29 Pronotal disc coarsely punctate; genae not projecting outward beyond eyes .... *Zophosis nigroaenea*

29' Pronotal disc glabrous; genae projecting outward beyond eyes .... *Zophosis punctata punctata*

30 Scutellar shield clearly visible 45

30' Scutellar shield invisible 31

31 Protibia slender, without any teeth; antenna with 11 distinct antennomers (Adesmiini) 32

31' Protibia provided with two large teeth on outer side; 11th antennomere very small, mostly hidden in 10th antennomere (Erodiini) 39

32 Metatibia strongly laterally compressed, blade-like 35

32' Metatibia round or oval, not sharp 33

33 Metatibia oval; elytral costa 2 almost reaches base, costa 3 (humeral costa) and 4 (lateral costa) are complete .... *Adesmia dilatata dilatata*

33' Metatibia round; elytra without distinct costa 34

34 Lateral crest distinctly shortened, only visible in the posterior half; sides of pronotum steeply sloping, almost angled towards the slightly concave disc .... *Adesmia cancellata abbreviata*

34' Lateral crest complete; pronotum normally rounded transversely .... *Adesmia montana montana*

35 Elytra long oval, disc convex .... *Adesmia cothurnata cothurnata*

35' Elytra short oval, disc flattened 36

36 On the lateral part of the elytra (pseudepipleuron) without or with very few microgranules .... *Adesmia metallica laevior*

36' On the lateral part of the elytra (pseudepipleuron) densely microgranulated in the half adjacent to the epipleuron .... *Adesmia metallica brozai*

37	Anterior leg with long hairs on tibia and femur (*Erodius*) 38
37'	Anterior leg without long hairs on tibia and femur .... *Amnodeis confluens confluens*
38	Elytra without humeral costa, or costa very weak 39
38'	Humeral costa on elytra present 40
39	Dorsal costa strongly shiny; somewhat more rounded shape .... *Erodius kneuckeri kneuckeri*
39'	Dorsal side including dorsal costa completely dull; shape a bit more elongate .... *Erodius opacus*
40	Only one dorsal costa on each elytra, located almost in the middle .... *Erodius zophosoides zophosoides*
40'	Two costae on each elytra or without a costa 41
41	Suture between the 10th and 11th antennomer straight, antennomer 11 only very slightly protruding from the tenth; pseudepipleural crest evenly curved from base to apex, apically not sinuate posteriorly; pronotum glabrous; costae high and shiny, inner dorsal costa about the same distance from suture as from outer dorsal costa .... *Erodius gibbus*
41'	Suture between the 10th and 11th antennomer curved, antennomer 11 relatively wide and pointed protruding from the tenth antennomer; pseudepipleural crest mostly almost straight in the middle, at least slightly sinuate posteriorly 42
42	Inner and outer dorsal costa high; pseudopleural crest evenly curved in the middle, slightly sinuate posteriorly .... *Erodius hebraicus*
42'	Only outer costa present, sometimes very weak or sometimes barely recognisable 43
43	Pronotum near posterior margin clearly punctured, traces of a dorsal costa barely visible; male with a tuft of bristles on prosternum .... *Erodius punctatus sinaiticus*
43'	Pronotum not punctured; at least outer costa present, male without a tuft of bristles on prosternum 44
44	Two inner dorsal costae are missing or only faintly indicated .... *Erodius dejeanii*
44'	Both inner dorsal costae and humeral costa about the same .... *Erodius duponchelii*
45	Sides of transverse pronotum distinctly curved upwards; anterior and posterior margins of pronotum very deeply cut, posterior corners pointed backwards in the males (Akidini) 46
45'	Sides of pronotum not strongly curved upwards 47

46	Elytra flat, both dorsal rows formed of granules .... *Akis reflexa*	
46'	Third interspace of elytra sloping obliquely, dorsal costa from base to elytral slope formed of few granules, at elytral slope as a complete costa .... *Akis spinosa*	
47	Antennae 11-segmented (Tentyriini) 50	
47'	Antennae 10-segmented (Adelostomini) 48	
48	Anterior margin of pronotum deeply, almost squarely emarginated, its posterior margin also deeply emarginated in the middle, posterior corners wing-like; tibial spur distinct, on protarsus longer than tarsomere 1; just one dorsal elytral costa near to the humeral costa .... *Machlopsis crenatocostata*	
48'	Anterior margin of pronotum not or weakly emarginated, its posterior margin also faintly; tibial spur indistinct, on protarsus much shorter than tarsomere 1; elytra with two dorsal costae: inner dorsal costa about as far from the suture as from the outer dorsal costa (*Adelostoma*) 49	
49	Sides of pronotum slightly rounded, not clearly heart-shaped. .... *Adelostoma sulcatum sulcatum*	
49'	Sides of pronotum apically strongly indented, heart-shaped (maybe a form of *A. sulcatum sulcatum*) .... *Adelostoma cordatum*	
50	Winged species; brown or brown-black; elytra long and narrow; shoulder angle and shoulder bump well developed 51	
50'	Unwinged species; black or brown; elytra usually shorter; shoulders often rounded, shoulder bump absent 53	
51	Upper side glabrous; base of elytra with fine margin .... *Cyphostehte heydeni*	
51'	Upper side with laid down light hairs; base of elytra unmargined 52	
52	Epistome truncate; tarsi glabrous above .... *Trichosphaena perraudierei perraudierei*	
52'	Epistome in the middle triangularly protruding; tarsi above and below clearly hairy; pubescence of the elytra patchy .... *Imatismus villosus*	
53	First tarsomer of protarsus lobed inwards; epistome projecting forwards, truncate .... *Prochoma audouini*	
53'	First tarsomer of protarsus normel, not lobed 54	
54	Epistome trilobed through a lateral cut-out on each side, the middle lobe large, with three small teeth, the mandibles completely exposed through the epistomal excision .... *Dailognatha crenata*	

54' Epistome never trilobed; the mandibles not completely exposed 55
55 Eye divided by tempus from back to front at least to the middle; base of elytra completely bordered 56
55' Eye simple, not divided from back to front 60
56 Elytra irregularly punctate; epistome thick, bead-like set off .... *Hionthis tentyrioides*
56' Elytra with regular rows of punctures, epistome normal (*Mesostena*) 57
57 Lower edge of eye seen clearly angled in lateral view 59
57' Lower edge of eye straight in lateral view 58
58 Anterior edge of epistome with a large asymmetrical tooth directed to the right .... *Mesostena puncticollis*
58' Anterior egde of epistome round, without a large asymmetrical tooth, but with numerous tiny teeth .... *Mesostena angustata angustata*
59 Third antennomer significantly longer than the second one .... *Mesostena peyerimhoffi*
59' Third antennomer significantly smaller than the second one .... *Mesostena picea sinaitica*
60 Pronotum closely joined to the elytra, almost always more narrowed towards the apex, or widest at the base 71
60' Pronotum only loosely joining the elytra, rarely narrowing to the apex more than to the base 61
61 Epipleural crest somewhat thickened and shaped file-like (*Oterophloeus*) 62
61' Epipleural crest smooth, not shaped file-like 63
62 Margin of the base of the pronotum reaching over the middle of each elytra; eye slightly convex .... *Oterophloeus alveatus peyerimhoffi*
62' Margin of base of pronotum shorter, occupying only about one-third of each elytra; eye distinctly convex .... *Oterophloeus haagi orientalis*
63 Protibia slender; apically not at all broadened outside 66
63' Protibia apically distinctly somewhat triangularly expanded outwards 64
64 Base of elytra not margined, only traces of a margin laterally sometimes visible; anterior tarsi short; antenna short, the third antennomere not twice

as long as the second; pronotum transverse, of the width of the elytra; brown .... *Scelosodis castaneum castaneum*

64' Base of elytra completely margined (*Schweinfurthia*) 65

65 Margins of pronotum complete, head with midline, outer spine of protibia as long as first 4 tarsomeres combined .... *Schweinfurthia sinaitica*

65' Margins of pronotum rudimentary, head without midline, outer spine of protibia shorter than first 4 tarsomeres combined, 7.5 mm .... *Schweinfurthia alfierii*

66 Head with clearly indicated eye fold; eyes convex or flat 68

66' Head without or only very faintly indicated eye fold; eyes convex (*Tentyrina*) 67

67 Pronotum wider than long, little convex .... *Tentyrina orbiculata subsulcata*

67' Pronotum as wide as long, more convex .... *Tentyrina palmeri palmeri*

68 Head (including slightly convex eyes) as wide as pronotum; antenna long, slender .... *Thraustocolus leptoderus*

68' Head (including eyes) narrower than pronotum (*Tentyria*) 69

69 Elytra with punctured rows, very finely confused punctuation between them .... *Tentyria punctatostriata*

69' Punctation of the elytra tangled 70

70 Eye conical; greatest width of pronotum in the middle .... *Tentyria discicollis*

70' Eye flat; greatest width of pronotum in front of the middle .... *Tentyria sinaitica*

71 Scutellar shield in front of the elytral basal line, which completely encloses it (*Micipsa*) 71

71' Tiny scutellar shield in or behind the elytral basal line, posteriorly always free, not enclosed 74

72 Elytra glabrous 73

72' Elytra with short erected hairs .... *Micipsa* spec.

73 Pronotum completely smooth, without longitudinal sculpture or punctation; eyes barely protruding from the lateral contour of the head .... *Micipsa douei douei*

73' Pronotum strongly longitudinally striated; eyes protruding far from the contours of the head .... *Micipsa schaumi*

74 Pronotum longitudinally arched, more narrowed towards the base than towards the anterior; tibia not sulcate on the outside .... *Hegeterocara arabica*

74' Pronotum flat, not longitudinally arched, more narrowed towards the anterior than towards the base; tibia sulcate on the outside (*Oxycara*) 75

75 Epipleural edge file-like channeled as a stridulatory organ .... *Oxycara peyerimhoffi*

75' Epipleural edge smooth 76

76 Sides of pronotum parallel, not converging distally .... *Oxycara ardoini*

76' Sides of pronotum not parallel, converging distally .... *Oxycara pygmaeum*

77 Integument covered with scales; sides of pronotum with large tooth, anterior margin bulged (Sepidiini) 82

77' Integument without scales; sides of pronotum without a tooth, anterior margin normal (Stenosini) 78

78 Eye completely divided by gena; the middle antennomeres wider than the basal and apical ones 79

78' Eye not completely divided by gena, gena just reaching the eye 80

79 Head, pronotum and elytra with costae .... *Microtelus careniceps binodiceps*

79' Head, pronotum and elytra without costae .... *Dichillus alfierii*

80 Ultimate antennomer smaller than the previous one; elytra with rows of punctures; no furrows on frons .... *Stenosis affinis*

81' Ultimate antennomer not smaller than the previous one; two deep furrows on frons .... *Mitotagenia aegyptiaca*

82 Antenna large, normally separated, not forming a club (*Sepidium*) 84

83' Antenna with solid, ovoid club ringed at the apex, 11th antennomer smaller and narrower than the 10th antennomer closely attached to it .... *Vieta tuberculata*

84 Shape robust; pronotum much wider than long; teeth on the sides of the elytra large, also the lateral pronotal bulb .... *Sepidium tricuspidatum tricuspidatum*

84' Shape gracile; pronotum much longer than wide; teeth on sides of elytra inconspicuous, the lateral bulb of pronotum small .... *Sepidium dathan*

KEY 3 IDENTIFICATION KEY FOR THE SUBFAMILIES BLAPTINAE AND TENEBRIONINAE

1 Epistome deeply emarginated, but not over the entire anterior edge of the epistome 13
1' Epistome rounded, truncate, if emarginated, then over the entire anterior edge of the epistome 2
2 Large beetle; antennomers 8 to 10 spherical, antennomer 11 conical (Blaptini: *Blaps*) 3
2' Smaller beetle, antenna formed differently 7
3 Caudal extension of the elytra lifted up in lateral view ....
   .... *Blaps bifurcata bifurcata*
3' Caudal extension of the elytra directed straight posteriorly ....
   4
4 Very large, 35–42 mm; elytra between the primary slightly ribbed stripes with fine secondary stripes, making the elytra appear double striped
   .... *Blaps wiedemanni*
4' Usually smaller than 35 mm, elytra not double striped ....
   5
5 Prosternal apophysis tightly fitting behind the procoxa, not forming a protruding edge; lateral margins of pronotum strongly and regularly rounded
   .... *Blaps judaeorum*
5' Prosternal apophysis forming a protruding edge behind the procoxa; lateral margins of pronotum not so much rounded outwards ....
   6
6 Mentum plane, not depressed; caudal extension of elytra longer ....
   .... *Blaps polychresta*
6' Mentum depressed anteriorly; caudal extension of elytra short ....
   .... *Blaps nitens laportei*
7 Body yellow-rufous, often semitransparent; slender; dorsally glabrous; pronotum not bordered; 6–7 mm (Helopini) ....
   .... *Hedyphanes chikatunovi*
7' At least lateral margins of pronotum bordered 8
8 Elytra with fine costae; shoulders rounded; head and pronotum wrinkled punctate; pronotum strongly rounded laterally, transversally arched, lateral margins not visible in dorsal view (Pedinini pars) ....
   .... *Drosochrus costatus aegyptiacus*
8' Pronotum transversally not arched, lateral margins visible in dorsal view
   9
9 All antennomers distinctly longer than wide, antennomere 11 much longer than the preceding ones; protibia with large tooth near the knee; >10 mm (Scaurini: *Scaurus*) 10

9' At least the middle antennomeres clearly wider than long; protibia with no tooth; <10 mm 11

10 Punctation of pronotum very uneven, on the disc fine and not very dense, near the lateral margins and at the base much more strongly punctate .... *Scaurus aegyptiacus*

10' Pronotum densely and strongly, almost evenly punctured .... *Scaurus rugicollis*

11 Larger, more than 5 mm; black; scutellar shield triangular .... *Alphitobius diaperinus*

11' Smaller, less than 3 mm; brown, scutellar shield pentagonal (Triboliini: *Tribolium*) 12

12 Antennae with well separated, 3-segmented club .... *Tribolium castaneum*

12' Antennae gradually widened distally, with ± set-off 5-6-segmented weak club .... *Tenebrio confusum*

13 Integument scaled; eye with scale-like bristles between the ommatids (Pedinini pars) .... *Leichenum pulcellum pumilum*

13' Integument not covered with scales; eye bare 14

14 Dorsal side smooth and shiny, glabrous, very weakly punctate, also the row punctation of the elytra very fine; epistome very deeply excised; posterior corner of pronotum rectangular; all tibiae on the outer margin armed with short spines, protibia distally dilated; eye only slightly constricted by gena; pro- and mesotasus in the male distinctly dilated, segments 1–4 strongly transverse .... *Diphyrrhynchus aenescens*

14' Characters not combined 15

15 Protibia of fossorial type: apically strongly widened distally or with pronounced teeth 26

15' Protibia not of fossorial type, slender, without any teeth 16

16 Eye completely divided by gena 17

16' Eye of gena constricted but not completely divided, some omatids remain visible at the narrowest part 18

17 Pronotum with rounded punctures, very fine on the disc, not confluent; the elytral intervals flattened and very finely granulated, the coriaceous ground covered with very short setae .... *Mesomorphus setosus*

17' Pronotum and elytra without setae; punctation of pronotum discally round, laterally more densely set, partly elongated .... *Opatroides punctulatus punctulatus*

18 Epipleuron ends at the level of the anterior edge of the abdominal ventrite V 20

18' Epipleuron extends to the end of the abdomen (Dendarini: *Dendarus*) 19

19 Pronotum wider, about 1.4 times as wide as long, lateral margin strongly and regularly rounded, largest width in the middle .... *Dendarus syriacus*

19' Pronotum narrower, about 1.3 times as wide as long, lateral margin weakly rounded, largest width in front of the middle .... *Dendarus calcaroides*

20 Pronotum with granules (Opatrini: *Gonocephalum*) 22

20' Pronotum with punctures (Opatrini: *Clitobius*) 21

21 Body black; wings fully developed .... *Clitobius oblongiusculus*

21' Body brown; wings reduced .... *Clitobius grimmi*

22 Small, length less than 5.5 mm; protibia relatively broad, about as broad as the first 4 tarsomeres combined; lateral margin of pronotum strongly sinuate apically .... *Gonocephalum setulosum*

22' Length more than 7 mm; protibia narrower, about as wide as the first two tarsomeres together, straight or almost straight; lateral margin of pronotum not or slightly sinuate apically 23

23 Large, length more than 12 mm; antennomer 3 about 3.5 times as long as wide .... *Gonocephalum soricinum soricinum*

23' Smaller than 10 mm; antennomer 3 about 3 times as long as wide 24

24 Lateral margin of pronotum apically not sinuate; elytra with two rows of setae per interstidium .... *Gonocephalum perplexum*

24' Lateral margin of pronotum apically slightly sinuate; elytra with three rows of hairs per interstidium 25

25 Base of pronotum with two mostly distinct oblique longitudinal impressions near the middle .... *Gonocephalum patruele patruele*

25' Base of pronotum without any longitudinal impression near the middle .... *Gonocephalum rusticum*

26 Protibia dilatated, but without strong teeth 30

26' Protibia armed with two strong teeth (Melanimonini: *Cheirodes*) 27

27 Epitome and frons with granules and setae .... *Cheirodes asperulus*

27' Epitome and frons not setigerous 28

28 Antenna 10-segmented; pronotum about 2.1 times as wide as long .... *Cheirodes brevicollis*

28' Antenna 11-segmented, pronotum less than 2 times as wide as long 29

29 Apical tooth of metatibia distally incised; epistome relatively shallowly excised .... *Cheirodes sardous sardous*

29' Apical tooth of metatibia distally not incised; epistome relatively very deeply excised .... *Cheirodes pilosus*

30 Posterior corner of pronotum completely rounded off; dorsal side bare and strongly punctured .... *Proscheimus arabicus*

30' Posterior corner of pronotum angled 31

31 Without furrow between epistome and frons; epipleural crest does not reach the apex of elytra 33

31' Epistome and frons separated by deep furrow; epipleural crest reaches apex of elytra 32

32 Lateral margin of elytra fimbriated with long hairs; interstitiae 3, 5 and 7 of elytra with narrow costa .... *Eurycaulus henoni*

32' Lateral margin of elytra not fimbriated; interstitiae of elytra convex but not costate .... *Sclerum orientale orientale*

33 Elongated; anterior margin of pronotum emarginated; elytra with rows of punctures; interstitiae with erect setae .... *Scleropatrum hirtulum*

33' Short oval, convex; anterior margin of pronotum weakly concave, almost straight; elytra without punctate rows, long, with pale erect setae; lateral margins of pronotum and elytra long pale ciliated .... *Ammobius rufus*

KEY 4   IDENTIFICATION KEY FOR THE SUBFAMILY DIAPERINAE

1 Protibia widened distally; elytra entirely glabrous 2

1' Protibia slender, not widened distally; elytra entirely pubescent .... *Pseudoseriscius maculosus murinus*

2 Antennae gradually widened 3

2' Antenna with six-segmented club, 11th segment small; elytra and pronotum laterally with long golden setae; body black .... *Trachyscelis tenuestriatus*

3 Yellow, often with a black macula; body elongated; eye slightly convex; anterior margin of pronotum narrower than the posterior margin; more than 5 mm .... *Phaleria prolixa*

3' Red brown; anterior margin of pronotum wider than the posterior margin; eye flat, not convex; less than 5 mm (genus *Phtora*) .... *Phtora subclavata / apicilaevis* [maybe synonyms]

KEY 5 IDENTIFICATION KEY FOR THE SUBFAMILY ALLECULINAE

1. First ventrite and metacoxae situated in the same level; first ventrite anteriorly, the others laterally margined; anterior edge of eye strongly emarginated by the genae; antennae inserted closely in front of eyes; terminal segments of maxillary palps much larger than preceding one ....(Alleculini) *Mycetocharina megalops megalops*

1' First ventrite lower than metacoxae; first ventrite anteriorly and the others laterally not bordered; anterior edge of eye not or only slightly margined, genae angles not touching eyes, antennae inserted below genae; terminal segment of the maxillary palps only slightly larger than the penultimate (Cteniopodini) 3

2. Epipleura developed only to metacoxae or reaching at most to third ventrite; pronotum narrower than elytra; body black, elytra brownish yellow .... *Omophlus gracilipes*

2' Epipleura almost reaching suture angle, but sometimes narrow behind .... *Cteniopus priesneri luteiceps*

VI Darkling Beetles Present in the Sinai Peninsula

In the text, the distribution indication "Egypt" refers to the African part of this country. The number of specimens examined is given in brackets. The system of families, subfamilies, tribes and species is according to Iwan et al. (2020) and Kamiński et al. (2020).

VI.1 Subfamily Pimeliinae Latreille, 1802

VI.1.1 Tribe Adelostomini Solier, 1834

VI.1.1.1 Genus *Adelostoma* Duponchel, 1827

VI.1.1.1.1 Adelostoma (Adelostoma) cordatum Solier, 1837

Adelostoma cordatum Solier, 1837: 169.

Referenced localities. The Sinai (Crotch 1872).

General distribution. North African chorotype: Libya (Gridelli 1930), Egypt (Alfieri 1976), the Sinai (Crotch 1872, Chikatunov et al. 1997), Israel (Koch 1935a).

Comments. The taxonomic status of *A. cordatum* is not clear. It is probably a variety of *A. sulcatum* (see Gebien 1937, Koch 1935a, Schawaller 1982).

VI.1.1.1.2 Adelostoma (Adelostoma) sulcatum sulcatum Duponchel, 1827 (see Plate I.1)

Adelostoma sulcatum Duponchel, 1827: 342.

Adelostoma sulcatum grandiforme Koch, 1935a: 39.

Material examined. Santa Katarina, 1.-2.XII.1992, L.P. (11); Santa Katarina, 16.IV.1995, P. (8); Santa Katarina, Wadi Tlach, 17.IV.1995 P. (8) (ML, TP).

Referenced localities. Wadi Isla (Alfieri 1976), Wadi Chamileh (Heyden 1899, Peyerimhoff 1907 t. Heyden).

General distribution. *A. sulcatum* s. str. Mediterraneo-Sindian chorotype: Morocco (Kocher 1958) up to Mesopotamia, southern Europe (Gridelli 1930), according to Kaszab (1981): on the European continent only in southern Spain. – *A. s. crassicorne* Peyerimhoff, 1931: 97: western, central and southern Sahara (Español 1973, Kocher 1958, Peyerimhoff 1931).

Comments. The specimens from the Sinai and Israel, available to us, measure between 5.5 and 8.5 mm. Koch (1935a) described *A. sulcatum grandiforme* as a race of *A. sulcatum* Duponchel. Size is, according to him, the only separation criterion. It should be between 5.0 and 6.5 mm for the Egyptian specimens of *A. sulcatum* s. str. and between 7 and 9 mm for *A. sulcatum grandiforme* from Israel. The non-available types of *A. sulcatum* Duponchel (Carl 1991a) measure between 7 and 9 mm (Duponchel 1827). No differentiation criteria on specimens from large parts of the distribution area could be found. *A. sulcatum grandiforme* Koch is therefore regarded as synonymous with *A. sulcatum* Duponchel (Lillig & Pavlíček 2003). Schawaller (1982) describes the variability of the pronotum of several varieties.

VI.1.1.2 Genus *Machlopsis* Pomel, 1871

VI.1.1.2.1 Machlopsis (Machlopsis) crenatocostata (Redtenbacher, 1868) (see Plate I.2)

Steira crenato-costata Redtenbacher, 1868: 120.
Referenced localities. Magdaba (Alfieri 1920, 1978).
General distribution. North African chorotype: Along the Mediterranean coast, from Tunisia to Syria (Gridelli 1930) and Israel (Renan et al. 2020), with the exception of southern Cyrenaica.
Comments. The species has been described as "vom Vorgebirge der guten Hoffnung" (Redtenbacher 1868). Haag-Rutenberg (1875a) could prove that this information was erroneous. *Machlopsis crenatocostata* was actively collected on the dunes in the middle of the night in September (Crovetti 1970). Koch (1937) found it in early summer.

VI.1.2 Tribe Adesmiini Lacordaire, 1859

VI.1.2.1 Genus *Adesmia* Fischer von Waldheim, 1822

VI.1.2.1.1 Adesmia (Adesmia) montana montana (Klug, 1830) (see Plate I.3)

Pimelia montana Klug, 1830: No. 25.
Adesmia perpolita Reitter, 1916a: 24.
Material examined. 7 km NE Santa Katarina, 1400 m, 2.XII.1992, L.P. (12); 16 km NE Santa Katarina, 1400 m, 2.XII.1992, L.P. (2); 22 km NE St. Katarina, 1300 m, in the garden, 2.XII.1992, L.P. (1); 27 km NE Santa Katarina, 1300 m, 2.XII.1992, L.P. (7); Wadi Rachaba, 1.III.1935, W. Wittmer (1); Fush el Arab, Kaiser, X.1926 (1) (MCSNM), Sinai, Ehrberg. (LT, 5 PLT) (MNHUB).
Referenced localities. Wadi Isla (Alfieri 1976), the Sinai (Andres 1931b sub *A. montana acervata* (Klug), Crotch 1872, Gridelli 1937, Kaszab 1981, Koch 1949, 1965, Peyerimhoff 1931, Reitter 1916a sub *A. perpolita*), Monte Sinai (Klug 1830), Mont Sinaï (Allard 1885), Jebel el Tih, Scheich Nebi Saleh (Ardoin 1978), Wadi Rachaba (Ardoin 1978 t. Koch), Wadi Ba'baa, Naqb el-Haoua, Qala'at el-Aqaba (Ardoin 1978 t. Peyerimhoff), Wadi es Scheich (Bodenheimer & Theodor 1929), Wadi Ba'baa (Heyden 1899), Wadi Rachaba (Koch 1935a), Wadi Ba'baa, Naqb el-Hasua [!], Qala'at el-Aqaba (Koch 1949 t. Peyerimhoff), Monti Sinai (Peyerimhoff 1907 t. Klug), Wadi Ba'baa (Peyerimhoff 1907 t. Heyden), Naqb el-Haoua, Qala'at el-Aqaba (Peyerimhoff 1907).
General distribution. SW-Asiatic chorotype: Southern Sinai, Jordan, Saudi Arabia (Kaszab 1981), southern Israel (Ardoin 1978), Iraq (Carl 1990).
Comments. Damage caused by this species on tomato seedlings has been observed in Jordan (Katbeh-Bader 1996). This species is significantly ($P = 0.041$) more prevalent in the South Sinai than in the North Sinai (Table II.1, Map VI.1).

Darkling Beetles Present in the Sinai Peninsula

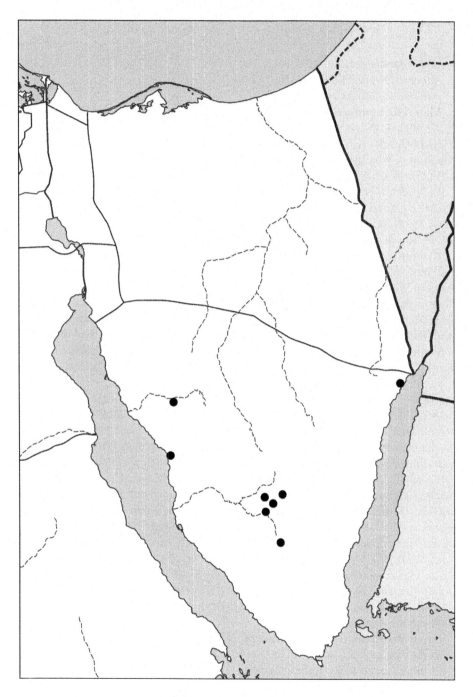

MAP VI.1 Distribution map of *Adesmia montana montana* in the Sinai Peninsula.

VI.1.2.1.2 Adesmia (Macradesmia) cancellata abbreviata (Klug, 1830) (see Plate I.4)

Pimelia abbreviata Klug, 1830: No. 27.
Adesmia latreillei Solier, 1835b: 540.
Adesmia sinaitica Crotch, 1872: 267.

Material examined. 20 km S Taba, 27.XI.1992, L.P. (1); 10 km N Nuweiba, 28.XI.1992, L.P. (1); 3 km N Nuweiba, 3.XII.1992, L.P. (1); 20 km S Nuweiba, 27.XI.1992, L.P. (1); Sainte Catherine Monastery, 1800 m, 1.XII.1992, L.P. (2); 27 km NE Sainte Catherine, 1300 m, 2.XII.1992, L.P. (5); 12 km NW Dahab, 30.XI.1992, L.P. (25); 18 km SW Dahab, 29.XI.1992, L.P. (2); 35 km SSW Dahab, 29.XI.1992, L.P. (3); 35 km N Sharm el Sheikh, 30.XI.1992, L.P. (7); Sinai 1929, leg. Jonarson (1) (MZH); Sinai occid., J. Couyat, III.1909 (2) (MNHNP); area around Ain Khudra oasis, 25.-29.XII.1997, 600–800 m, leg. W. Schawaller (1) (SMNS), ca. 750 m, 17 km SW of Nuweiba, Zirig Wadi, 5.IV.1996, leg. Gy. M. László, (2) (TMB).

Referenced localities. Wadi Karam, El Arish, Magdaba (Alfieri 1976), Meghara and Wadi Mezara (Alfieri 1920, 1976 sub *A. cancellata*), south of Ain Musa, between Wadi Feiran and Wadi Selaf, between Wadi Hebran, Ga'a desert and Tor, Wadi Gergir, Wadi Budr, south of Wadi Chaschibi, between Wadi Tarfa and central Sinai Massif (Andres 1920), everywhere on the Sinai Peninsula (Andres 1931b), Wadi Kid, Wadi Kunei-el-E-Rayah near Dahab, Abas Bosha near of the Monastery of Sainte Catherine, Bir Zreier, Wadi Habeik near Bir Zreier, Wadi Niddya el Beida near Firan Sinai, Wadi Ba'baa near Umm Bugma, Ma'agama, Ein Chadijeh, 10 km S Nakhl, Wadi Rayane, Sarabit el Hadam, Ein Zaituna, Surgit, El Hadan Aridi (Ardoin 1978), Wadi Abutli, Wadi Rachaba, Wadi Cheikh, Wadi Hebran (Ardoin 1978 t. Koch), Djebel Hamr, Djebel Sefrou, Djebel el Hih, Wadi Gneh, Djebel Serbal, Deir el-Arbain, Wadi Tayebeh (Ardoin 1978 t. Peyerimhoff), Wadi Isla (Bodenheimer & Theodor 1929 sub *A. sinaitica*), Sinai (Boehm 1908 sub *A. clathrata cancellata*, Crotch 1872 sub *A. clathrata*, *A. sinaitica*, Gridelli 1953, Reitter 1916a sub *A. cancellata*), Djebel Hamr, Djebel Sefrou, Djebel el Tih (Heyden 1899 sub *A. cancellata*), Wadi Abutli, Wadi Rachaba, Wadi Feiran, Wadi Isla, Wadi Cheikh, Wadi Hebran (Koch 1935a sub *A. sinaitica*, Koch 1951), Wadi Gneh, Djebel Serbal, Deir el-Arbain, Wadi Tayebeh (Peyerimhoff 1907 sub *A. clathrata*), Djebel Hamr, Djebel Sefrou, Djebel el Tih (Peyerimhoff 1907 t. Heyden sub *A. cancellata*), Wadi Ba'baa (Peyerimhoff 1907 t. Heyden sub *A. clathrata*), Djebel Hamr, Djebel Lefrou, Djebel el-Tik, Quady Gneh, Djebel Serrar, Deir-el-Arbain, Quady Tangebeh (Koch 1951 t. Peyerimhoff), El Arbain, El Haschue, Hmed, Wadi Gorondel, high plateau of Ga'a (Koch 1951), St. Katherine Protectorate (Zalat et al. 2008 sub *A. sinaitica*).

General distribution. *A. cancellata abbreviata*: Endemic to the Sinai. – *A. cancellata cancellata* (Klug, 1830, No 37): Israel, Syria (Ardoin 1978), Arabian Peninsula (Kaszab 1981), Jordan, Pakistan (Iwan et al. 2020). – *A. cancellata clathrata* Solier 1835: Iran, Iraq, Oman, Saudi Arabia, Syria (Iwan et al. 2020). – *A. cancellata hadramautina* Kaszab, 1981: Yemen.

Comments. The type comparison between *Adesmia latreillei* Solier and *Pimelia abbreviata* Klug showed the identity of both taxa. Carl (1990) noticed the similarity between *A. abbreviata* and *A. cancellata* and declared them as synonymous. He underlined the weak sculpture of the elytra. This characteristic distinguishes the samples from the Sinai that Ardoin (1978) classified as subspecies of *A. cancellata*. The examination of large collections showed the large variability of all the related taxa. The distinction according to the criteria named by Reitter (1916a) and Koch (1951) is not always possible.

In Lillig and Pavlíček (2003) *A. cancellata abbreviata* was named *A. cancellata latreillei*. This species is significantly ($P = 0.003$) more prevalent in the South Sinai than in the North Sinai (Table II.1, Map VI.2).

Pimelia abbreviata Klug, 1830: Lectotype (designated by Lillig & Pavlíček 2003): Male. 14693/Hist.-Coll. (Coleoptera) Nr. 14693, *Adesmia abbreviata* Klug, Syria, Ehrbrg., Zool. Mus. Berlin/*abbreviata* Klug* (*Pimelia*), Syria. Ehrbg./*Adesmia cancellata* Klug, det. M. Carl, 1990.

Paralectotype: Syria. Ehrbg./Hist.-Coll. Nr. 14693/Hist.-Coll. (Coleoptera) Nr. 14693, *Adesmia abbreviata* Klug, Syria, Ehrbrg., Zool. Mus. Berlin/Zool. Mus. Berlin/*Adesmia cancellata* Klug, det. M. Carl, 1990, 1 male (MNHUB).

Adesmia latreillei Solier, 1835: Lectotype (designated by Lillig & Pavlíček 2003): Male. 1605, 95/*Adesmia latreillei* Dej. Paralectotype: 18, 34, 1 Ex. (MNHNP).

All specimens of *A. cancellata cancellata* (Klug, 1830) recorded from the Sinai (Map VI.2) are to be related to *A. cancellata latreillei*. The specimens from southern Israel that we examined are transition forms of *A. cancellata cancellata* and *A. cancellata latreillei*. This observation confirms Ardoin's (1978) opinion.

VI.1.2.1.3 Adesmia (Oteroscelis) cothurnata cothurnata (Forskål, 1775) (see Plate II.1)

Tenebrio cothurnatus Forskål, 1775: 80.

Pimelia bicarinata Klug, 1830: No. 31 (syn. according to Lillig & Pavlíček 2003).

Adesmia subserrata Chevrolat, 1877: 113 (syn. according to Lillig & Pavlíček 2003).

Adesmia cothurnata glabrior Schatzmayr & Koch, 1934: 23 (syn. according to Lillig & Pavlíček 2003).

Adesmia cothurnata omanensis Kaszab, 1981: 353 (syn. according to Lillig & Pavlíček 2003).

Material examined. 7 km S Taba, 3.XII.1992, L.P. (1); 13 km S Taba, 3.XII.1992, L.P. (1); 16 km S Taba, 27.XI.1992, L.P. (5); 19 km S Taba, 3.XII.1992, L.P. (9); 38 km S Taba, 3.XII.1992, L.P. (1); 22 km N Nuweiba 3.XII.1992, L.P. (3); 10 km N Nuweiba, 28.XI.1992, L.P. (2); 3 km N Nuweiba, 3.XII.1992, L.P. (11); 20 km NNW Dahab, 2.XII.1992, L.P. (1); 12 km NW Dahab, 30.XI.1992, L.P. (1); 7 km NE Santa Katarina, 1400 m, 2.XII.1992, L.P. (15); Santa Katarina, 1500 m, 1.-2.XII.1992, L.P. (4); 1–6 km E Santa Katarina, 8.IV.1993, P. (3); 27 km NE Santa Katarina, 1300 m, 2.XII.1992, L.P. (10); 45 km SSW Dahab, 29.XI.1992, L.P. (1); 35 km N Sharm el Sheikh, 30.XI.1992, L.P. (8); 15 km N Sharm el Sheikh, 30.XI.1992, L.P. (6); 7 km N Sharm el Sheik, 29.XI.1992, L.P. (1); Santa Katarina, Plain of Raha, 17.4.1995, P. (6); El Arish, 19.XII.1947, leg. Henry Field (3); SW Sinai, 10.I.1948, leg. Henry Field (4) (BM); Sharm el Sheikh, 26.XII.1980, leg. R. Miravalle (8) (SB); ET – Wadi

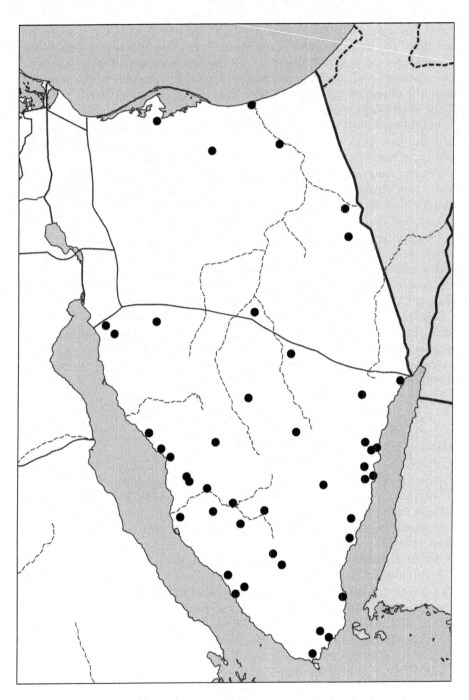

MAP VI.2 Distribution map of *Adesmia cancellata abbreviata* in the Sinai Peninsula.

Feiran, 28.45N 33.20E, 04.1993, leg. Ullrich (3); ET – Nuweiba, 28.58N 34.38E, 04.1993, leg. Ullrich (6) (CGW); Wadi Isla, 27.2.<19>35, leg. W. Wittmer (1 PT *glabrior*); Wadi Racha, 3.3.<19>35, leg. W. Wittmer (4 PT *glabrior*); Wadi Hebran, 6.3.19>35, leg. W. Wittmer (1 PT *glabrior*); Tor, Deserto verso Isla, 27.2.<19>33, leg. A. Schatzmayr (2 PT *glabrior*) (MCSNM); Sharm el Sheik around Shark's Bay, 1.-4.I.1998, leg. W. Schawaller (3) (SMNS); 17 km SW of Nuweiba, Wadi Zirig, ca. 750 m, 5.IV.1996., leg. Gy. M. Lázló (1) (TMB); Sinai, Oasis Ain Hudra, 19.4.2000, 28°55'N 34°33'E, Leg. Schmid-Egger (3).

Referenced localities. Wadi Isla (sub *A. bicarinata bicarinata*), Karm Alam, Wadi Meghara, Cheikh Hamid (sub *A. bicarinata glabrior*) (Alfieri 1976), south from Ain Musa up to Wadi Gharandel, between Wadi Schellal and Wadi Mokatteb, between Wadi Feiran and Wadi Selaf, Wadi Hebran, Desert of Ga'a, Tor, Wadi Gergir, Wadi Chaschibi, south of Wadi Chaschibi, Wadi Tarfa (Andres 1920), the Sinai (Andres 1931b sub *A. bicarinata*, Crotch 1872 sub *A. bicarinata*, Gridelli 1937 sub *A. bicarinata*, Gridelli 1953 sub *A. cothurnata glabrior*, Hafez & Makki 1959 sub *A. bicarinata*, Kaszab 1979 sub *A. c. glabrior*, Koch 1940b sub *A. cothurnata glabrior* Schatzmayr & Koch, Reitter 1916a sub *A. bicarinata*), Ma'agana, Wadi Kid, Charm el Sheik, Monastery of St. Catherine, Wadi Abu-Ga'da, 40 km S Abu Rhodes, Bir Abu Rudais (Ardoin 1978 sub *Adesmia cothurnata glabrior*), Tor, Wadi Isla (Ardoin 1978 t. Schatzmayr & Koch), Wadi Isla, Wadi Hebran, Wadi Cheikh, Wadi Racha, Wadi Rachaba, Wadi Feiran (Ardoin 1978 t. Koch), Wadi Chebeikeh, Wadi Ba'baa, Wadi Khamileh, Wadi Fairan, Wadi El Arish (Ardoin 1978 t. Peyerimhoff), Wadi Shebeke, Wadi Ba'baa, Wadi Chamile, Wadi Bel Maï, Wadi Feiran, Wadi El Arish (Heyden 1899 sub *A. bicarinata*), between Wadi Tarfa and Sinai Massif, northern part of the Sinai Peninsula (Kneucker 1903), Wadi Isla, Wadi Hebran, Wadi Cheikh, Wadi Racha, Wadi Rachaba, Wadi Feiran (Koch 1935a sub *A. bicarinata glabrior* Schatzmayr & Koch, 1934), Wadi Chebeikeh, Wadi Ba'baa, Wadi Khamileh, Wadi Fairan, Wadi El Arish (Peyerimhoff 1907 sub *A. bicarinata*), between Wadi Tarfa and Sinai Massif, Naqb Boudra, Naqb el-Haoua, Hadjar er-Reqab, Wadi Gneh, Wadi Feiran, Wadi Gharandel, Wadi Tayebeh, Ain el-Houdra (Peyerimhoff 1907), between Tor and Wadi Isla (Schatzmayr & Koch 1934 sub *A. bicarinata glabrior* sp. n.).

General distribution. NE-African-Sindian chorotype: Egypt (locus typicus: Alexandria), the Sinai, southern Israel, southern Jordan, Saudi Arabia, Bahrain, Yemen, United Arab Emirates, Oman, Iraq, Sudan.

Comments. The criteria, above all the sculpture of elytra, hind tibiae and outer part of mandibles, given by Andres (1920), Koch (1935a) and Reitter (1916a) to differentiate between the supposed Upper Egyptian *Adesmia cothurnata* and the Lower Egyptian *A. bicarinata*, are unusable. The sculpture of elytra varies all over the distribution area. In the Sinai, Lower and Middle Egypt, it is possible to find specimens whose inner side of the hind tibiae displays ridges (*bicarinata*) as well as some specimens with hind tibiae without ridges (*cothurnata*). All studied specimens which originated from areas located in the south of Edfu (an Egyptian city located on the western side of the Nile river) do not show ridges on their tibiae. The mandible's outer face is usually, however not always, without ridges (*cothurnata*) in the northern part of the distribution area and the elytra show increasingly marked ridges towards the south. These facts indicate that the above-mentioned criteria are only clinal attributes which are not appropriate for the differentiation of subspecies. *Adesmia cothurnata* has been frequently mentioned for Upper Egypt and excluded from Lower Egypt (cf. Koch 1940b). Forskål (1775) described the species as "Kahirae in desertis frequens, in urbe repertus rarior". Klug (1830) mentioned Alexandria as a collection site for the species – which is probably incorrect. Secure evidences exist only from the right bank of the Nile. Although the type specimens of *A. cothurnata* had been destroyed around 1800 (Carl 1990), the short description of the problems does not justify any doubt regarding the identity of the species with the type series of *A. bicarinata* (MNHUB). That is to say *A. bicarinata* should be regarded as a synonym of *A. cothurnata*.

Gebien (1937) expressed doubts as to *A. subserrata* of being a synonym of *A. lacunosa* (Klug, 1830). A female of the type series (MNHNP) at our disposal shows the synonymy to *A. cothurnata*.

At our disposal are also the holotype (BM) and nine paratypes (one PT from BM, eight from MCSNM) of *A. glabrior*. According to Schatzmayr & Koch (1934), *A. glabrior* is distinct from *A. bicarinata* (= *cothurnata*) due to following characteristics:

- a smoother, far less sculptured upper side,
- the rows of tubercles composed of rather pointed grains, that are usually to be found in the intercoastal spaces, have mostly completely disappeared,
- the false epipleura of the elytra mostly show no grains or only some very isolated grains.

Koch (1935a) distinguishes one form from another in the following way:

- elytra almost smooth, without clear tubercle sculpture for *glabrior*,
- strongly marked tubercles between the external dorsal costa and the lateral costa are missing in *glabrior*.

All these characteristics examined by us on more than 200 specimens originating mainly from Egypt and Israel are extremely variable.

Kaszab (1981) describes *A. cothurnata omanensis*, according to four specimens at his disposal, of which we saw two (TMB). In our opinion, these and numerous other specimens from Oman and the United Arab Emirates (HJB, ML, ONHM) belong to the variation spectrum of *A. cothurnata* s. str.

This species is significantly ($P = 2.10^{-4}$) more prevalent at the South Sinai than in the North Sinai (Table II.1, Map VI.3).

VI.1.2.1.4   *Adesmia (Oteroscelis) metallica brozai* Ardoin, 1978 (see Plate II.2)

Adesmia metallica brozai Ardoin, 1978: 300.

Material examined. Between El Arish and Mohamdiya, X-XII.1916, leg. Major E. E. Austen (1) (BM); Sebchat A-Sheck (NE El Arish) (1) (ML).

General distribution. Southern Levantine chorotype: NW Negev (Ardoin 1978), NE Sinai.

Darkling Beetles Present in the Sinai Peninsula

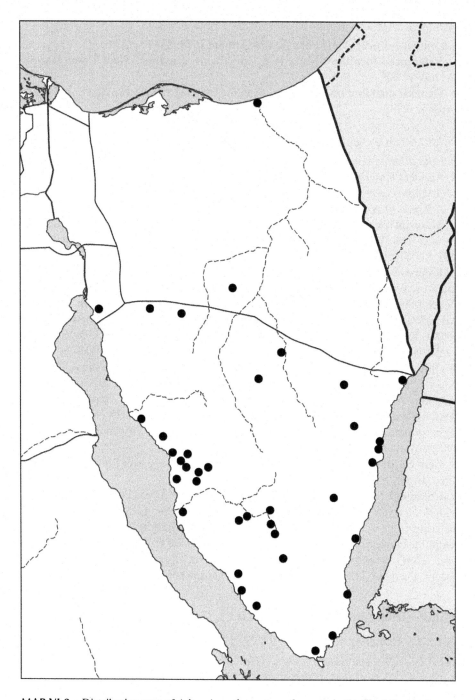

MAP VI.3 Distribution map of *Adesmia cothurnata cothurnata* in the Sinai Peninsula.

VI.1.2.1.5 Adesmia (Oteroscelis) metallica laevior Ardoin, 1978 (see Plate II.3)
Adesmia (*Oteroscelis*) *metallica laevior* Ardoin, 1978: 298.
Referenced localities. Mitla Pass, Bir Tmade, Kuseima, Wadi Umm Matirdi (Ardoin 1978).
General distribution. Southern Levantine chorotype: Northern Sinai, the Negev (Ardoin 1978).

VI.1.2.1.6 Adesmia (Oteroscelopsis) dilatata dilatata (Klug, 1830) (see Plate II.4)
Pimelia dilatata Klug, 1830: No. 32.
Adesmia macropus Solier, 1835b: 542.
Adesmia convergens Walker, 1871: 14.
Adesmia drakii Crotch, 1872: 267 (syn. restit. by Lillig & Pavlíček 2003).
Material examined. G. M'rara, 16.I.<19>70, Zinner/Sinai/C 15008/Muséum Paris ex coll. R. Oberthur (1) (NMHNP); El Arish, 19.XII.1947, leg. Henry Field (19); SW Sinai, 10.1.1947, leg. H. Field (1) (BM).
Referenced localities. Wadi Helal, Cheikh Hamid, El Kreig (Alfieri 1920, 1976), south of Ain Musa (Andres 1920), El Arish (Andres 1931b sub *A. drakei* Crotch), Gebel M'rara (Ardoin 1978 sub *A. drakei drakei*), Hammam Fir'aoun, Wadi El Arish, Wadi Gharandel (Ardoin 1978 t. Peyerimhoff), El Arish, Aium Mussa (Ardoin 1978 t. Koch), Sinai (Crotch 1872 sub *A. drakei*, Gridelli 1930, 1937, Peyerimhoff 1935 t. *A. dilatata* var. *drakei*, Schuster 1934 sub *A. drakei drakei*), Wadi el Arish, Wadi Bel Maï Heyden 1899 sub *A. macropus*), El Arish (Koch 1935a sub *A. dilatata drakei* Crotch, Koch 1940b), Aioum Mussa (Koch 1940b t. Andres 1920), Hammam Fir'aoun, Wadi Gharandel (Koch 1940b t. Peyerimhoff 1907), Wadi Chebeikeh, Wadi Ba'baa, Wadi Khamileh, Wadi Fairan (Peyerimhoff 1907); Wadi El Arish (Peyerimhoff 1907 t. Heyden), Pharaoh's Baths (= Hammam Fir'aoun) (Peyerimhoff 1907 t. Walker sub *A. convergens* sp. n.).
General distribution. Southern Levantine chorotype: Egypt (Alfieri 1976), the Sinai, Israel, Jordan (Ardoin 1978).
Comments. *Adesmia drakii* Crotch was recognised by Blair (1935) as synonymous with *A. dilatata*. Koch (1940b) regards the specimens from the Sinai as a subspecies differing from the nominal form and that is characterised by a reduced superficial sculpture. However, the examination of large series (BM, HJB, ML, MCSNM, MZH, TAU, ZSM) showed that this characteristic varies in all regions in which *A. dilatata* occurs. Therefore, *A. drakei* should be regarded as synonymous with *A. dilatata*.

VI.1.3 Tribe Akidini Billberg, 1820

VI.1.3.1 Genus *Akis* Herbst, 1799

VI.1.3.1.1 Akis spinosa (Linnaeus, 1764) (see Plate III.1)
Tenebrio spinosa Linnaeus, 1764: 101.
Akis spinosa var. *sculptior* Koch, 1935: 51.
Material examined. Wadi Nasb, 20 km ESE Monastery of Sainte-Catherine, 1.X.1981, leg. R. Summkeller (1); 40 km NNW Dahab, 500 m, 2.XII.1992, L.P. (1); 27 km NE Sainte-Catherine, 1400 m, 2.XII.1992, L.P. (2) (ML).

Referenced localities. Wadi Isla (Alfieri 1976), the Sinai (Crotch 1872, Gridelli 1933a, b, Gridelli 1953), Ain Musa (Hart 1891, Peyerimhoff 1907 t. Hart), Wadi Isla (Alfieri 1976).

General distribution. Arabian chorotype: Jordan, Arabian Peninsula (Soldati 2009), Israel (Negev) (ML). Iwan et al. (2020) also mention Algeria (Hoggar), Libya, Syria, Iraq. At least the North African entries are questionable.

Comments. Published by Lillig & Pavlíček (2003) as *A. elevata* var. *sculptior* Koch, 1935. Ferrer (2008) examined the holotype of *Tenebrio spinosus* Linnaeus and found this species to be identical to *A. elevata* var. *sculptior* Koch.

It is possible that records of *A. elevata* (including Iwan et al. 2020) refer to *A. elevata* var. *sculptior*.

The specimen from 40 km NNW Dahab was found in a sandy wadi, the one reported from the proximity of Monastery of Sainte-Catherine was found in a gravel desert.

VI.1.3.1.2 Akis reflexa (Fabricius, 1775) (see Plate III.2)
Pimelia reflexa Fabricius, 1775: 252.
Referenced localities. Magdaba (Alfieri 1920), Magdaba, Rafa (Alfieri 1976), Náchle (Heyden 1899), El Arish (Koch 1935a), Qala'at en Nakhel (Peyerimhoff 1907 t. Heyden), the Sinai (Schawaller 1987).

General distribution. Levantine chorotype: Lower Egypt, the Sinai (Schawaller 1987), Israel (Negev) (ML), and Syria (Iwan et al. 2020).

Comments. To be found in the coastal area mostly under stones and in the inland part mostly in houses (Koch 1935a).

VI.1.4 Tribe Erodiini Billberg, 1820

VI.1.4.1 Genus *Amnodeis* Miller, 1858

VI.1.4.1.1 Amnodeis confluens confluens Miller, 1858 (see Plate III.3)
Amnodeis confluens Miller, 1858: 123.
Referenced localities. The Sinai (Kaszab 1981).

General distribution. *Amnodeis confluens confluens*. SW-Asiatic chorotype: Iraq, Jordan, the Sinai (Kaszab 1981), the Negev (Bytinski-Salz 1955a). – *A. c. freyi* Schuster, 1937: Syria.

Comments. Some of the Israeli specimens (TAU) are labelled: *Amnodeis gebieni* aut *confluens*? det. Gridelli 1949. Some of the mentioned specimens correspond to the type series of *gebieni* Reitter, 1914 (GF), whereas others have the characteristics attributed by Reitter (1914a) to *A. confluens* Miller, 1858. There are many transitions between them. *A. gebieni* seems thus to be synonymous with *A. confluens*. A formal synonymisation is not appropriate here since we did not study the Miller types.

VI.1.4.2 Genus *Erodius* Fabricius, 1775

VI.1.4.2.1 Erodius (Dimeriseis) kneuckeri kneuckeri Andres, 1920 (see Plate III.4)
Erodius kneuckeri Andres, 1920: 61.
Erodius bilineatus Boehm, 1909: 121 nec Olivier, 1791: 426.

Material examined. Sebchat A-Sheck (NE El Arish), 10.V.1980 (8) (ML, TMB); El Arish, 12 km N of, 27.XII.1956, O. Yarkoni (1); El-Arish, P. Amitai, 11.I.<19>57 (1); Rodmano, 20.VI.1970, leg. Bytinski-Salz (2); 35 km NW Refidemn(?) 21.III.<19>76, leg. D. Gersting (3); Mitla Pass, 13.IV.1973, leg. Bytinski-Salz (5); Mitle, 13.IV.1973, leg. D. Furth (8); Mitla, 13.IV.1973, leg. M. Kaplan (1) (TAU).

Referenced localities. Gebel Katherin (Alfieri 1976), between Ain Musa and Wadi Werdan, between Hawara, Wadi Gharandel and Wadi Uset (Andres 1920), the Sinai (Andres 1931b).

General distribution. Southern Levantine chorotype: *E. kneuckeri kneuckeri*: Lower Egypt (Ismailiya) (Koch 1935a), the Sinai (Alfieri 1976), the Negev (TAU, TMB). – *E. kneuckeri semisculptus* Koch, 1935: Lower Egypt (Alfieri 1976).

VI.1.4.2.2 Erodius (Dimeriseis) opacus Kraatz, 1865

Erodius opacus Kraatz, 1865: 25.

Referenced localities. Lanafet Rissan, Magdaba (Alfieri 1976), the Sinai (Crotch 1872).

General distribution. Southern Levantine chorotype: Egypt (Oasis of Kharga), the Sinai (Alfieri 1976), Israel (Ayal & Merkl 1994).

VI.1.4.2.3 Erodius (Dirosis) dejeanii Solier, 1834 (see Plate IV.1)

Erodius dejeanii Solier, 1834: 543.

Referenced localities. Wadi Isla, Wadi Karm Alam (Alfieri 1976), Pharaoh's Baths (= Hammam Fir'aoun) (Walker 1871, determination uncertain).

General distribution. Levantine chorotype: Israel (Koch 1935b), the Sinai (Alfieri 1976), Syria (Iwan et al. 2020).

Comments. According to Gebien (1937), the species occurs also in Mesopotamia which is very unlikely.

VI.1.4.2.4 Erodius (Dirosis) duponchelii Allard, 1865

Erodius (Dirosis) duponchelii Allard, 1865: 388.

 E. dejeani var. *lineolatus* Reitter, 1914: 73.

Referenced localities. Wadi Isla (Alfieri 1976 sub *E. dejeani* var. *lineolatus* Reitter), El Arish (Andres 1931b, Koch 1935a sub *E. dejeanii duponcheli* (Allard)).

General distribution. Levantine chorotype: Israel (Koch 1935b), the Sinai (Alfieri 1976), Syria (Iwan et al. 2020).

VI.1.4.2.5 Erodius (Dirosis) hebraicus Lillig, 1997 (see Plate IV.2)

Erodius (Dirosis) hebraicus Lillig, 1997: 55.

Referenced localities. Mitle, 13.IV.1973, D. Furth (1 PT) (Lillig 1997).

General distribution. Southern Levantine chorotype: Northern Sinai, Israel (the Negev) (Lillig 1997).

VI.1.4.2.6 Erodius (Dirosis) puncticollis sinaiticus Crotch, 1872 (see Plate IV.3)

Erodius sinaiticus Crotch, 1872: 267.

 Erodius maillei auct. nec Solier, 1834: 546.

Erodius puncticollis var. *contractus* Kraatz, 1865: 65.
Referenced localities. Between Abu Diab and El Mishedi, between Wadi Mezara and Um Shusha (Alfieri 1920 sub *E. puncticollis contractus*, Alfieri 1976 sub *E. puncticollis* Solier), Wadi Isla, Karm Alam (Alfieri 1976 sub *E. puncticollis* var. *maillei*), south of Wadi Musa up to Wadi Werdan (Andres 1920 sub *E. puncticollis*), in north-west (Andres 1920 sub *E. puncticollis* var. *contractus*), Wadi Werdan, Wadi Werdan – Ain Hawara, Wadi Hebran, desert plain Ga'a up to Tor, Djebel Hamam, Wadi Gergir, Wadi Ta-albi, between Wadi Chaschibi, Djebel Nimr and Wadi Ab-Orta (Andres 1920 sub *E. puncticollis* var. *maillei*), the Sinai (Andres 1931b sub *E. maillei* and *E. contractus*, Crotch 1872 sub *E. puncticollis*, Gridelli 1933a sub *E. puncticollis* var. *maillei*, *E. sinaiticus*, Pierre 1961b sub *E. sinaiticus*, Reitter 1914a sub *E. puncticollis* var. *maillei*), El Arish (Boyd 1917), northern part of the Sinai Peninsula (Kneucker 1903), between Tor and Wadi Isla (Koch 1935a), Abu Rueitat (Koch 1940a).
General distribution. Southern Levantine chorotype: The Sinai (Koch 1940a), the Negev (ML). – *E. puncticollis puncticollis* Solier, 1834: Egypt, Libya (Pierre 1961b). – *E. puncticollis mirei* Pierre, 1961: Chad (Tibesti).
Comments. The records published as *Erodius servillei* Solier, 1834 by Andres (1931b) and Peyerimhoff (1907) should perhaps be added here. *E. puncticollis sinaiticus* is significantly ($p = 0.027$) more prevalent in South Sinai than in North Sinai (Table II.1, Map VI.4).

VI.1.4.2.7 Erodius (Erodius) gibbus gibbus Fabricius, 1775 (see Plate IV.4)
Erodius gibbus Fabricius, 1775: 258.
Referenced localities. El Arish (Andres 1931b), Wadi Gharandel (Walker 1871, determination uncertain for Walker, Peyerimhoff 1907 t. Walker).
General distribution. Levantine chorotype: *E. gibbus gibbus*: Soloum in Egypt (Alfieri 1976) up to northern Israel (Koch 1935b). – *E. gibbus cyrenaicus* Schuster, 1926: Libya (Cyrenaica).
Comments. *E. gibbus* is distributed mostly in coastal regions, but can penetrate sandy zones in the hinterland. *E. cyrenaicus* from Agedabia (Libya) was described as good species but lately has been classified as a subspecies of *E. gibbus* by Koch (1937).

VI.1.4.2.8 Erodius (Zophoserodius) zophosoides zophosoides Allard, 1865 (see Plate V.1)
Erodius zophosoides Allard, 1865: 387.
Material examined. Wadi Gharandel, IV.1993, leg. Dr. G. W. Ullrich (1) (HJB).
Referenced localities. Meghara (Alfieri 1976).
General distribution. *E. zophosoides zophosoides*: North African chorotype: Algeria, Tunisia, Libya, Egypt (Andres 1931b), the Sinai. – *E. zophosoides schatzmayri* Koch, 1937: Libya (Tripolitania) (Koch 1937). – *E. zophosoides nunicus* Kocher, 1955: southern Morocco (Kocher 1958). – *Erodius zophosoides marrakensis* Kocher, 1950: Morocco (Marrakesh) (Kocher 1958). – *E. zophosoides subbicostatus* Koch, 1937: Libya (Agedabia) (Koch 1937).

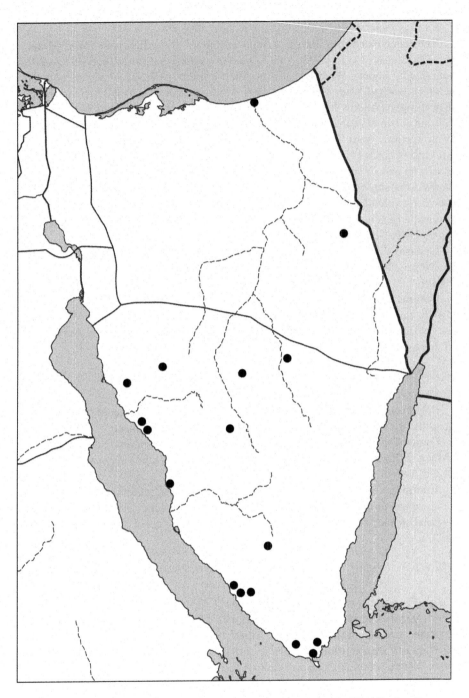

MAP VI.4   Distribution map of *Erodius puncticollis sinaiticus* in the Sinai Peninsula.

VI.1.5 Tribe Pimeliini Latreille, 1802

VI.1.5.1 Genus *Astorthocnemis* Lillig & Pavlíček, 2002

VI.1.5.1.1 Astorthocnemis becvarorum Lillig & Pavlíček, 2002 (see Plate V.2)
Astorthocnemis becvarorum Lillig & Pavlíček, 2002: 99.
Referenced localities. Dahab (Lillig & Pavlíček 2002).
General distribution. Southern Levantine chorotype: Jordan (vicinity of Aqaba), the Sinai (Dahab) (Lillig & Pavlíček 2002).
Comments. Only one specimen from the Sinai was found under a stone in the screecovered desert (Lillig & Pavlíček 2002) in contrast to the fact that the long tarsi covered with bristles lead one to suppose a sabulicole way of life (Pierre 1958). The closely related genus *Storthocnemis* is to be found mostly in sandy habitats and only seldom found in regs and hamadas (Pierre 1961a).

VI.1.5.2 Genus *Pimelia* Fabricius, 1775

VI.1.5.2.1 Pimelia (Pimelia) angulata angulata Fabricius, 1775 (see Plate V.3)
Pimelia angulata Fabricius, 1775: 252.
  Pimelia alternata Klug, 1830: No 16.
  Pimelia aculeata Klug, 1830: No 17.
  Pimelia angulata sinaitica Schatzmayr & Koch, 1934: 26.
Material examined. Wadi Gharandel, IV.1993, leg. G. W. Ullrich (30) (HJB).
Referenced localities. Wadi Helal, Meghara (Alfieri 1920, 1976), Ain Musa, Wadi Werdan, Wadi Gharandel, Wadi Feiran, Wadi Hebran, Tor (Andres 1920), Wadi Hebran (Andres 1920 sub *P. angulata* var. *aculeata*), Ain Musa, Wadi Werdan (Andres 1920 sub *P. angulata* var. *alternata*), El Arish (Andres 1929a), Wadi Gharandel, Wadi El Arish (Andres 1931b t. Heyden sub *P. nilotica* var. *aegyptiaca* Andres), El Arish, Tor (Andres 1931b), between Suez and Abu Selima (Bodenheimer & Theodor 1929), Romani (Boyd 1917), the Sinai (Crotch 1872, Gridelli 1930 t. Andres), Wadi Gharandel, Wadi Werdan, Wadi Ba'baa, Wadi Bei Maï (Heyden 1899), between Wadi Werdan and Suez (Kneucker 1903), El Arish (Koch 1935a, Koch 1940b t. Andres 1929a), Wadi Ouerdan, Wadi Ba'baa (Peyerimhoff 1907 t. Heyden), Wadi Gharandel, Hadjar er-Reqab, desert of Schur (Peyerimhoff 1907), Ain Musa, Mount Sinai (Walker 1871 sub *P. aculeata*), the Sinai (Innes Bey 1912); in wadis (Alfieri 1976), Tor (Koch 1935a, 1940b, Schatzmayr & Koch 1934); Abu Rheital, Bir Gifgafa, El Arich, Gebel el Maghara, Sharm el Sheikh, Shark Bay, Tor (Leo 2020).
General distribution. Levantine chorotype: North-eastern Egypt, Israel, northern Sinai (Koch 1940b). – *Pimelia angulata nilotica* Sénac, 1884: Upper Egypt, Sudan, Chad, Niger (Koch 1940b, Leo 2020). – *Pimelia angulata angulosa* A. G. Olivier, 1795: Mali, Mauritania, Guinea, Senegal (Leo 2020).
Comments. Leo (2020) revised the *Pimelia angulata* taxa group. He proposed to treat *P. angulata sinaitica* as a junior synonym of the nominate subspecies, just like *P. angulata syriaca* Sénac, 1884. Furthermore, he treats *Pimelia angulata tschadensis* Koch, 1940 as a junior synonym of *Pimelia angulata nilotica* Sénac, 1884 and *Pimelia angulata berlandi* Ardoin, 1963 as a junior synonym of *Pimelia angulata*

angulosa A. G. Olivier, 1795. These changes could no longer be considered by Iwan et al. (2020).

VI.1.5.2.2 Pimelia (Pimelia) arabica edomita Koch, 1940 (see Plate V.4)

Pimelia (Homalopus) arabica edomita Koch, 1940a: 102.

Material examined. Nuweiba, 27.III.1996, leg. N. Rech (1); Nuweiba, 28.XI.1992, L.P. (6); 40 km NNW Dahab, 2.XII.1992, L.P. (2); 20 km NNW Dahab, 2.XII.1992, L.P. (1); 55 km NE Santa Katarina, 2.XII.1992, L.P. (1) (ML); area around oasis Ain Khudra, 25.-29.XII.1997, 600-800 m, leg. W. Schawaller (2) (SMNS); Abu Zenima, 2.X.1992, leg. Ullrich (6) (CGW, ML).

Referenced localities. The Sinai (Alfieri 1976 t. Koch), in the north-west of the Sinai Peninsula (Andres 1920), Ain Musa (Andres 1920 sub *P. arabica* var. n.), Ain el Houdra, Wadi Feiran etc. (Andres 1931b sub *P. arabica* Klug), Ain el-Houdra, Wadi Feiran (Gridelli 1953 sub *P. arabica* t. Peyerimhoff, Peyerimhoff 1907 sub *P. arabica*), northern part of the Sinai Peninsula (Kneucker 1903), Wadi Feiran (Koch 1935a sub *P. arabica* Klug), Wadi Feiran, Ain Houdrah (Koch 1940a sub *P. arabica*, Gridelli 1953 t. Koch, Peyerimhoff), Wadi Feiran, Ain Hondrah (Kaszab 1982 t. Koch), northern part of the peninsula (Peyerimhoff 1907 t. Kneucker sub *P. arabica*).

General distribution. Southern Levantine chorotype: *P. a. edomita*: the Sinai, southern Jordan (Kaszab 1982), the Negev (ML). – *P. a. arabica* Klug, 1830 No 18: Saudi Arabia, North Yemen, Iraq (Kaszab 1982), according to Iwan et al. (2020) also in United Arab Emirates, Bahrain, Jordan, Oman. – *P. a. omanica* Kaszab, 1982: 148: Oman, United Arab Emirates (Kaszab 1982). – *P. a. emiri* Kaszab, 1982: 149: United Arab Emirates, Bahrain, North Yemen (Kaszab 1982).

Comments. Kaszab (1982) is not sure whether to regard *P. arabica omanica* and *P. arabica emiri* as good species or subspecies. The examination of specimens from Oman (ML, ONHM) leads one to suppose that these are only varieties. In addition, Kwieton (1982) names also *P. thomasi* Blair, 1931 as infra-specific form of the *P. arabica* complex. Kaszab (1982) lists them as an individual polytypic species. Specimens from the Sinai, which Koch (1940a) calls *P. arabica arabica* can be distinguished from *P. arabica edomita* of Wadi Araba (ZSM) by the characteristics given by Koch (1940a): sutural interstice more densely and more prominently covered with granules, inner dorsal costa reaching the base of elytron, costa higher at the declining apical part of the elytra, the suture forming a costa on each side. The genuine *P. arabica arabica* Klug, 1830 from the Arabian Peninsula is even more sculptured. The taxa group is in the need of revision.

VI.1.5.2.3 Pimelia (Pimelia) barthelemyi Solier, 1836 (see Plate VI.1)

Pimelia barthelemyi Solier, 1836b: 350.

Referenced localities. Magdaba (Alfieri 1920), Wadi Ain El Gederat (Alfieri 1976), Ain Musa, desert plain of Ga'a, between Gergir and Wadi Budr (Andres 1920), the Sinai (Andres 1931b, Gridelli 1937), El Qantara, Romani, Mohammedia (Boyd 1917), Magdaba (Gridelli 1930 t. Alfieri, Andres), Tor (Koch 1935a), desert of Schur, Hadjar er-Reqab (Peyerimhoff 1907).

General distribution. North African chorotype: Libya (Cyrenaica), Egypt, the Sinai,? Syria (Gridelli 1937).

Comment. Not present in Syria (Iwan et al. 2020).

VI.1.5.2.4 *Pimelia (Pimelia) bottae* Sénac, 1887

Pimelia bottae Sénac, 1887: 48.

Referenced localities. El Arish, Wadi Karam (Alfieri 1976), Wadi El Arish (Andres 1931b, Peyerimhoff 1907 t. Heyden sub *P. bottai* Sénac), El Arish (Gridelli 1953 t. Peyerimhoff 1907, Koch 1935a), Wadi El Arish, Wadi Bel Maï (Heyden 1899).

General distribution. Arabian chorotype: Saudi Arabia, North Yemen (Kaszab 1982), Qatar (Mas-Peinado et al. 2018), the Sinai (Kaszab 1979, Koch 1935a, Peyerimhoff 1907).

VI.1.5.2.5 *Pimelia (Pimelia) canescens canescens* Klug, 1830 (see Plate VI.2)

Pimelia canescens Klug, 1830: No 3.

Referenced localities. Magdaba (Alfieri 1920, 1976), the Sinai (Andres 1931b), Magdaba (Gridelli 1930 t. Alfieri).

General distribution. *P. canescens canescens*: North African chorotype: Mediterranean coast of Egypt, the Sinai (Alfieri 1976), Israel (Ayal & Merkl 1994), Libya (Iwan et al. 2020). – *P. canescens arabicola* Schatzmayr & Koch, 1934: Lower Egypt (Helouan) (Alfieri 1976).

VI.1.5.2.6 *Pimelia (Pimelia) hirtella* Sénac, 1887

Pimelia hirtella Sénac, 1887: 47.

Referenced localities. Wadi Lagama (Alfieri 1976), the Sinai (Andres 1931b).

General distribution. SW-Asiatic chorotype: The Sinai (Andres 1931b, Alfieri 1976), Egypt, Saudi Arabia (Kaszab 1982).

VI.1.5.2.7 *Pimelia (Pimelia) mittrei* Solier, 1836 (see Plate VI.3)

Pimelia mittrei Solier, 1836a: 134.

Referenced localities. Wadi Lagama, Wadi Helal (Alfieri 1976), El Arish (Andres 1931b, Koch 1935a).

General distribution. Levantine chorotype: Israel (Koch 1935b), the Sinai (Koch 1935a), Lebanon (Iwan et al. 2020).

Comments. Solier (1836a) describes *P. mittrei* as "De Grèce ... et d'Égypte", but it is for sure not present in Greece (cf. Kühnelt 1965).

VI.1.5.2.8 *Pimelia (Pimelia) subquadrata subquadrata* Sturm, 1826 (see Plate VI.4)

Pimelia subquadrata Sturm, 1826: 68.

Pimelia irrorata Solier, 1836a: 99.

Material examined. Bir Gafgafa, 11.III.1968, Broza & al. (1) (ML).

Referenced localities. Wadi Ain El Gederat, Wadi Isla (Alfieri 1976), south of Ain Musa (Andres 1920), Wadi Chamileh (Andres 1920 t. Heyden), between Suez and Abu Selima (Bodenheimer & Theodor 1929 sub *P. irrorata*), Mohammedia (Boyd 1917 sub *P. irrorata*), the Sinai (Gridelli 1930), Wadi Chamile (Heyden 1899), Wadi Khamileh (Peyerimhoff 1907 t. Heyden).

General distribution. Levantine chorotype: Palestine, the Sinai, Egypt (Gridelli 1930).

VI.1.5.2.9   Pimelia (Pimelia) theveneti Sénac, 1880
Pimelia theveneti Sénac, 1880: 267.
Referenced localities. El Kontella, Bir Isla (Alfieri 1976).
General distribution. Southern Levantine chorotype: Egypt, the Sinai (Alfieri 1976).

VI.1.5.3   Genus *Prionotheca* Dejean, 1834

VI.1.5.3.1   Prionotheca coronata coronata (Olivier, 1795) (see Plate VII.1)
Pimelia coronata Olivier, 1795, 59: 4.
Referenced localities. Meghara (Alfieri 1920, 1976), between Ain Musa and Wadi Werdan, desert plain of Ga'a (Andres 1920), the Sinai (Andres 1931b, Gridelli 1930 t. Alfieri and Andres, Gridelli 1933b, 1937, 1950, 1952, 1953, Español 1943, 1967, Kaszab 1979, Koch 1934, Schatzmayr 1938, Peyerimhoff 1931), between Suez and Abu Selima (Bodenheimer & Theodor 1929), El Kubri (Boyd 1917), Wadi Gharandel (Hart 1885, 1891), Wadi El Arish, Wadi Gharandel, Wadi Sidr (Heyden 1899), Ain Musa, Wadi Gharandel, Wadi Sidreh, Wadi El Arish (Peyerimhoff 1907 t. Walker, Hart, Heyden), Ain Musa, Berg Sinai (Walker 1871).
General distribution. *P. coronata coronata*. Saharo-Sahelo-Arabian chorotype: North Africa, the Sinai (Koch 1934), Jordan (vid. Lillig). – *P. coronata freyi* Bytinski-Salz, 1955b: 314: the Negev. – *P. coronata ovalis* Ancey, 1881: 397: Arabian Peninsula, Iran, Iraq (Kaszab 1982).
Comments. The transition between *Prionotheca coronata coronata* and *P. coronata freyi* in the Sinai could be observed near Nod el Naza (Bytinski-Salz 1955b).

VI.1.5.4   Genus *Pterolasia* Solier, 1836

VI.1.5.4.1   Pterolasia squalida Solier, 1836 (see Plate VII.2)
Pterolasia squalida Solier, 1836a: 68.
Thriptera lanata Peyerimhoff, 1907: 35.
Referenced localities. Wadi Werdan, between Tor and Djebel Hamam (Andres 1920 sub *Thriptera lanata*), the Sinai (Andres 1921, Ardoin 1963, 1971b, Español 1967), El Arish (Andres 1927b, 1931b, Gridelli 1930, 1952), Hadjar er Regab, Haserith (Gridelli 1952 t. Peyerimhoff 1907, Innes Bey 1909 t. Peyerimhoff, Peyerimhoff 1907 sub *Thriptera lanata* sp. n.), the Sinai (Gridelli 1953).
General distribution. Saharian chorotype: Disjunct in western Senegal and former Spanish Sahara, in eastern Libya (Cyrenaica), Egypt, the Sinai (Español 1967), the Negev (Bytinski-Salz 1955a).
Comments. Buried under bushes of *Zilla spinosa* (L.) Prantl in the wadis near Heluan (part of Cairo, Lower Egypt) (Andres 1920, 1927b), under *Hyoscyamus muticus* L. on the beach of El Arish (Andres 1927b).

VI.1.5.5   Genus *Thriptera* Solier, 1836

VI.1.5.5.1   Thriptera crinita crinita (Klug, 1830)
Pimelia crinita Klug, 1830: No 22.
Referenced localities. Wadi Isla (Alfieri 1976), the Sinai (Andres 1929b, 1931b, Gridelli 1940 t. Andres, Gridelli 1950, 1954, Koch 1934, Pierre 1961b), Wadi Isla, Wadi Hebran (Koch 1935a).

General distribution. NE-African-Sindian chorotype: Egypt, Sudan, Eritrea, Jordan, Saudi Arabia, Yemen (Kaszab 1982), Iran (Iwan et al. 2020), the Sinai (Alfieri 1976). – *T. crinita gilleti* Pierre, 1979: Sudan (Jebel Marra).

VI.1.5.5.2   Thriptera kraatzi Haag-Rutenberg, 1876 (see Plate VII.3)

Thriptera kraatzi Haag-Rutenberg, 1876b: 75.
   Thriptera sorocula Reitter, 1894: 303.
   Material examined. 8 km S Taba, 26.XI.1992, L.P. (1); 14 km S Taba, 27.XI.1992, L.P. (2); 19 km S Taba (1), 3.XII.1992, L.P.; 3 km N Nuweiba, 3.XII.1992, L.P. (3); 20 km NW Dahab, 2.XII.1992, L.P. (1); 12 km NW Dahab, 30.XI.1992, L.P. (1); 1 km W Dahab, 29.XI.1992, L.P. (2); Dahab, 22.4.1995, P. (1) (ML, TP); Nuweiba, 28°58'N 34°38'E, 27.VI.1994, leg. Ullrich (17) (CGW).
   Referenced localities. Gulf of Aqaba (Alfieri 1976, Andres 1931b t. Peyerimhoff), north of the Gulf of Aqaba (Peyerimhoff 1907 sub *Th. sorocula*), the Sinai (Ferrer 1995b).
   General distribution. NE-African-Sindian chorotype: Egypt, Ethiopia, India (Rajastan), Iran, Jordan, Oman, Pakistan, Saudi Arabia, Sinai, Somalia, Sudan, Yemen (Ferrer 1995b, Iwan et al. 2020, Kwieton 1978, 1981).
   Comments. Unlike other authors (among others Kwieton 1978), Andres (1929b) and Koch (1941b) regard the synonymy between *Thriptera kraatzi* and *Thriptera sorocula* as unjustified.

VI.1.5.5.3   Thriptera varvasi pilipes Kraatz, 1865

Thriptera pilipes Kraatz, 1865: 299.
   Thriptera heydeni Reitter, 1893: 214.
   Thriptera guyoti Andres, 1920: 70.
   Referenced localities. Qantara, Wadi Helal (Alfieri 1976), the Sinai (Andres 1920 sub *Th. guyoti* sp. n., Andres 1929b, 1931b), Mezar, El Arish (Boyd 1917 sub *Th. varvasi* Solier), Hadjar er-Reqab (Peyerimhoff 1907 sub *Th. heydeni*).
   General distribution. *Th. varvasi varvasi* Solier, 1836: 52: Moroccan-Algerian border up to western Cyrenaica (Peyerimhoff 1948a). – *Th. varvasi pilipes*: Southern Levantine chorotype: Lower Egypt up to the Sinai (Alfieri 1976, Andres 1931).

VI.1.5.6   Genus *Trachyderma* Latreille, 1828

VI.1.5.6.1   Trachyderma (Trachyderma) andresi (Gridelli, 1928)

Ocnera andresi Gridelli, 1928: 16.
   Referenced localities. Magdaba, Wadi Isla (Alfieri 1976 sub *Ocnera andresi*), Deir el Arbein (Andres 1928, 1931b, Gridelli 1928, 1930), the Sinai (Gridelli 1954).
   General distribution. Endemic to the Sinai (Gridelli 1954).
   Comments. This species seems to be taxonomically still unclear (cf. Gridelli 1953, Koch 1935a).

VI.1.5.6.2   Trachyderma (Trachyderma) genei Solier, 1836

Trachyderma genei Solier, 1836a: 58.
   Referenced localities. Wadi Lagama, Wadi Isla (Alfieri 1976 sub *Ocnera genei*), the Sinai (Constantinou & Cloudsley-Thompson 1983, St Katherine Protectorate, V. - VI. 2010 (Norfolk et al. 2012).

General distribution. North African chorotype: Libya (Cyrenaica), Egypt (Gridelli 1954), the Sinai (Alfieri 1976).

VI.1.5.6.3 Trachyderma (Trachyderma) hispidum (Forskål, 1775) (see Plate VII.4)
Tenebrio hispidus Forskål, 1775: 79.
Material examined. 8 km S Taba, 26.XI.1992, L.P. (1); 14 km S Taba, 27.XI.1992, L.P. (1); 19 km S Taba, 3.XII.1992, L.P. (3); Nuweiba, 28.XI.1992, L.P. (5); 10 km NW Dahab, 2.XII.1992, L.P. (1); Dahab, 28.XI.1992, L.P. (2); 35 km N Sharm el Sheikh, 30.XI.1992, L.P. (1); 15 km N Sharm el Sheikh, 30.XI.1992, L.P. (7); Sharm el Sheikh, 29.-30.XI.1992, L.P. (3) (ML).
Referenced localities. Cheik Hamid, El Kreig, Magdaba, Wadi Ain El Gederat (Alfieri 1920, 1976 sub *Ocnera hispida*), Wadi Isla (Alfieri 1976 sub *Ocnera hispida*), between Ain Musa and Wadi Werdan, Oasis of Feiran, Wadi Hebran, desert plain of Ga'a, Tor, between Tor and Djebel Hamam, between Wadi Gergir and Wadi Budr, Wadi Chaschibi, south of Wadi Chaschibi, northern part of the peninsula (Andres 1920 sub *Ocnera hispida*), between Suez and Abu Selima (Bodenheimer & Theodor 1929 sub *Ocnera hispida*), the Sinai (Crotch 1872 sub *Ocnera hispida, O. gomorrhana*, Koch 1934 sub *Ocnera hispida latreillei* Solier), Wadi el Ain (Hart 1891 sub *Ocnera hispida*), Nachle, Wadi Sidr, Wadi El Arish (Heyden 1899 sub *Ocnera hispida*), northern part of the Sinai Peninsula (Kneucker 1903 sub *Ocnera hispida*), Tor, Wadi Feiran (Koch 1935a sub *Ocnera hispida latreillei* Solier), Ain Musa (Peyerimhoff 1907 t. Walker), Wadi Sidreh, Wadi El Arish, Qala'at en Nakhel (Peyerimhoff 1907 t. Heyden), Hadjar er-Reqab, Wadi Gneh, Wadi el-Ain, in the north of the Peninsula (Peyerimhoff 1907 sub *Ocnera hispida*), Ain Musa, Berg Sinai (Walker 1871 sub *Ocnera hispida*), St. Katherine Protectorate (Zalat et al. 2008 sub *Ocnera hispida* Forskål).
General distribution. Saharo-Sahelo-Arabian chorotype: North Africa from Morocco to Egypt, Sicily, Sudan, Near East, Arabian Peninsula (Kaszab 1982).
Comments. According to Koch (1935b), *Trachyderma hispidum gomorrhanum* Reiche & Saulcy, 1857: 215 differs from *T. hispidum latreillei* Solier, 1836a: 36 through a small prominent hump on the bent part of the prosternal apophysis and through broader and shorter elytra. The examination of 150 specimens of *T. hispidum* s. l. from the whole distribution area showed that both criteria are variable. The somewhat prominent hump is to be found in some specimens from Ein Gedi (Israel) (locus classicus of *T. gomorrhanus* is Gomorrha near the Dead Sea) and from Inzinkat, Niger (ML). No characteristics are to be established that enable an unquestionable distinction between *T. hispidum gomorrhanum* and *T. hispidum latreillei*. Both forms seem to be identical. Formal synonymisation is not possible here since the taxonomic problems are not yet satisfactorily solved (cf. Peyerimhoff 1948a). Skopin (1962) establishes the genus *Atrachyderma* for the *Trachyderma* because of the different structures of male femora and male genitalia. While Medvedev & Nepesova (1985) consider *Atrachyderma* a genus, Kwieton (1981) lists it as a subgenus of *Trachyderma*. Kaszab (1982) mentions *Atrachyderma* neither as a genus or as a subgenus.
The lethal temperature of this nocturnal species is 45°C at 10 percent humidity for 24-hour exposure. Its subelytral cavity reduces water loss from the body and is thus an adaptation mechanism to the arid habitat (Cloudsley-Thompson 1965).

VI.1.5.6.4 Trachyderma (Trachyderma) philistinum Reiche & Saulcy, 1857 (see Plate VIII.1)

Trachyderma philistina Reiche & Saulcy, 1857: 214.
?Ocnera parvicollis Baudi di Selve, 1875: 686.
Ocnera habelmanni Kraatz, 1865: 292.
Material examined. 27 km NE Santa Katarina, 1400 m, 2.XII.1992, L.P. (1); Santa Katarina, 16.IV.1995, P. (1) (ML); Sharm el Sheik around Shark's Bay, 1.-4.I.1998, leg. W. Schawaller (4) (SMNS).
Referenced localities. Wadi Karam (Alfieri 1920 sub *Ocnera hispida*), Karm Allam, El Arish (Alfieri 1976 sub *O. parvicollis*), Wadi Helal, Wadi Karam, Wadi Isla (Alfieri 1976 sub *Ocnera philistina*), Mont Sinai (Allard 1883 sub *Ocnera philistina*), between Ain Musa and Wadi Werdan, Oase Feiran, between Tor and Djebel Hamam, Tor (Andres 1920 sub *Ocnera philistina*), between Tor and Djebel Hamam, Wadi Gergir, Wadi Chaschibi (Andres 1920 sub *Ocnera habelmanni*), the Sinai (Andres 1921 sub *Ocnera habelmanni*), Andres 1928, 1931b sub *Ocnera parvicollis*, Constantinou & Cloudsley-Thompson 1983, Schuster 1938), Wadi Feiran (Heyden 1899 sub *Ocnera habelmanni*, Koch 1935a sub *Ocnera philistina* var., Peyerimhoff 1907 sub *O. habelmanni* t. Heyden, Gridelli 1928 t. Peyerimhoff 1907), Hadjar er-Reqab, Wadi Feiran (Peyerimhoff 1907 sub *Ocnera philistina*), St Katherine Protectorate (Zalat et al. 2008 sub *Ocnera philistina* Reiche).
General distribution. Mediterraneo-Sindian chorotype: Rhodes, Turkey ("Asia Minor"), Cyprus, Israel, Syria, Egypt, Jordan, Iran, Iraq, Arabia, India (Grimm 1991).
Comments. According to Soldati (2009) the real *T. parvicollis* Baudi di Selve, 1875 is distributed in Iran and Qatar.

VI.1.6 Tribe Sepidiini Eschscholtz, 1829

VI.1.6.1 Genus *Sepidium* Fabricius, 1775

VI.1.6.1.1 Sepidium dathan Crotch, 1872 (see Plate VIII.2)

Sepidium dathan Crotch, 1872: 268.
Sepidium abiram Crotch, 1872: 268.
Material examined. 7 km NE Santa Katarina, 1400 m, 2.XII.1992, L.P. (2) (ML).
Referenced localities. Wadi Ain El Gederat, Gebel El Arish (Alfieri 1976), the Sinai (Crotch 1872 sub *Sepidium dathan* n. sp., *S. abiram* n. sp., Reitter 1914b t. Crotch).
General distribution. East Mediterranean chorotype: Libya (Cyrenaica), Egypt, Lebanon (Gridelli 1930), Israel (ML), Syria (Andres 1931b).

VI.1.6.1.2 Sepidium tricuspidatum tricuspidatum Fabricius, 1775 (see Plate VIII.3)

Sepidium tricuspidatum Fabricius, 1775: 250.
Sepidium korah Crotch, 1872: 268.
Referenced localities. Wadi Isla (Alfieri 1976), the Sinai (Crotch 1872 sub *S. tricuspidatum* and sub *S. korah*), Wadi El Arish (Peyerimhoff 1907 t. Heyden), Wadi

Gneh (Peyerimhoff 1907), Wadi Gaifi (Andres 1931b, Koch 1935a t. Andres sub *S. korah*).

General distribution. North African chorotype: Libya (Cyrenaica) up to Syria (Kwieton 1980), the Negev (Bytinski-Salz 1955a), Gaza (Reitter 1914b), four other subspecies in North Africa and Malta (Kwieton 1980).

VI.1.6.2 Genus *Vieta* Laporte de Castelnau, 1840

VI.1.6.2.1 Vieta tuberculata (Solier, 1844)

Dymonus tuberculatus Solier, 1844: 223.
Referenced localities. The Sinai (Kaszab 1982).
General distribution. Arabian chorotype: Egypt, the Sinai, Saudi Arabia, Yemen (Kaszab 1982).

VI.1.7 Tribe Stenosini Schaum, 1859 (1934)

VI.1.7.1 Genus *Dichillus* Jacquelin du Val, 1860

VI.1.7.1.1 Dichillus (Dichillus) alfierii Koch, 1935 (see Plate VIII.4)

Dichillus alfierii Koch, 1935a: 46.
Material examined. Mt. Moses, 16.IV.1995, P. (1).
Referenced localities. Wadi Cheikh (Koch 1935a).
General distribution. Endemic to the Sinai (Koch 1935a).

VI.1.7.2 Genus *Microtelus* Solier, 1838

VI.1.7.2.1 Microtelus careniceps binodiceps Reitter, 1907

Microtelus binodiceps Reitter, 1907: 115.
Microtelus careniceps sinaiticus Koch, 1935a: 48.
Referenced localities. Wadi Helal, Wadi Isla (Alfieri 1976 sub *M. careniceps careniceps* = *M. binodiceps*), between Wadi Werdan and Wadi Feiran, Oase Feiran (Andres 1920 sub *M. careniceps*), Wadi El Tarfa (Koch 1935a sub *M. careniceps sinaiticus* ssp. n.), the Sinai (Holdhaus 1919 sub *M. careniceps*, Kaszab 1982 sub *M. careniceps sinaiticus*).
General distribution. *M. careniceps careniceps* Reiche & Saulcy, 1857: 227: Lebanon, Israel, South-West Jordan (Carl 1992a), Egypt, Iraq, Syria (Iwan et al. 2020).

M. careniceps binodiceps: Southern Levantine chorotype: Lower Egypt, the Sinai (Carl 1992a).
Comments. Carl (1992a,b) mentions hybridisations between *Microtelus careniceps* s. l. and *M. lethierryi* Reiche, 1860.

VI.1.7.3 Genus *Mitotagenia* Reitter, 1916

VI.1.7.3.1 Mitotagenia aegyptiaca Koch, 1941

Mitotagenia aegyptiaca Koch, 1941a: 40.
Mitotagenia arabs auct. nec (Baudi di Selve, 1881: 279).
Referenced localities. Wadi Isla (Alfieri 1976), the Sinai (Andres 1931a, Kaszab 1979, 1982), Auel Habesh (Andres 1931b sub *M. arabs*, Koch 1941a t. Andres), Wadi Feiran, Aoul Haibeh, Wadi Sa'al (Peyerimhoff 1907 sub *Eutagenia arabs*).

General distribution. SW-Asiatic chorotype: Egypt (Nile valley), the Sinai, Saudi Arabia (Kaszab 1982), Israel (the Negev, Chikatunov in litt.).

Comments. This species was wrongly reported as *M. arabs* (Baudi di Selve, 1881) which is to be found only in Yemen and Ethiopia (Koch 1941a).

VI.1.7.4 Genus *Stenosis* Herbst, 1799

VI.1.7.4.1 Stenosis affinis (Solier, 1838)

Tagenia affinis Solier, 1838a: 26.

Referenced localities. Wadi Isla (Alfieri 1976).

General distribution. Southern Levantine chorotype: Lower Egypt, the Sinai (Alfieri 1976)

VI.1.8 Tribe Tentyriini Eschscholtz, 1831

VI.1.8.1 Genus *Cyphostethe* Marseul, 1867

VI.1.8.1.1 Cyphostethe (Cyphostethoides) heydeni (Haag-Rutenberg, 1877) (see Plate IX.1)

Himatismus heydeni Haag-Rutenberg, 1877: 282.
Himatismus saharensis Chobaut, 1897: 205.

Referenced localities. Wadi Meghara (Alfieri 1976), between Suez and Abu Selima (Bodenheimer & Theodor 1929 sub *Curimosphena heydeni*).

General distribution. Saharo-Sahelo-Arabian chorotype: From Sahara, across and beyond its southern border, up to Sudan (Koch 1950), Saudi Arabia, Yemen (Kaszab 1981).

VI.1.8.2 Genus *Dailognatha* Steven, 1828

VI.1.8.2.1 Dailognatha crenata Reiche & Saulcy, 1857 (see Plate IX.2)

Dailognatha crenata Reiche & Saulcy, 1857: 198.

Referenced localities. Wadi Meghara (Alfieri 1976), El Arish (Koch 1935a).

General distribution. SW-Asiatic chorotype: Cyprus, Iran, Syria, Israel, Jordan, the Sinai, Lebanon, Turkey (Iwan et al. 2020).

VI.1.8.3 Genus *Hegeterocara* Reitter, 1900

VI.1.8.3.1 Hegeterocara arabica Reitter, 1900 (see Plate IX.3)

Hegeterocara arabica Reitter, 1900: 190.
Fourtaus brevicornis Pic, 1921: 48.

Material examined. Nuweiba, IV.1992, Ullrich (1) (CGW); 7 km S Taba, 3.XII.1992, L.P. (1); 38 km S Taba, 3.XII.1992, L.P. (1); 22 km N Nuweiba, 3.XII.1992, L.P. (1) (ML).

Referenced localities. The Sinai (Pic 1920 sub *Fourtaus brevicornis* n. sp.), Cheikh Hamid (Alfieri 1976), Wadi Helal (Alfieri 1976 sub *Fourtaus brevicornis*), El Arish, Gebel el Helal (Koch 1935a), Sharm el Sheik (Typus, Reitter 1900), El Kontella (Koch 1941b).

General distribution. Southern Levantine chorotype: The Sinai, southern Jordan (Kaszab 1981), the Negev (TAU).

VI.1.8.4 Genus *Hionthis* Miller, 1861

VI.1.8.4.1 Hionthis tentyrioides Miller, 1861 (see Plate IX.4)
Hionthis tentyrioides Miller, 1861: 176.
Referenced localities. Karm Alarm (Alfieri 1976), Wadi Feiran (Ardoin 1972a).
General distribution. East Mediterranean chorotype: Syria (Schawaller 1982), Libya (Cyrenaica), Lower Egypt (Gridelli 1930), Palestine (Bodenheimer 1937), Israel (Negev) (Ayal & Merkl 1994).

VI.1.8.5 Genus *Imatismus* Dejean 1834

VI.1.8.5.1 Imatismus (Imatismus) villosus (Haag-Rutenberg, 1870) (see Plate X.1)
Himatismus villosus Haag-Rutenberg, 1870: 90.
Material examined. Dahab, 20.IV.1995, P. (1) (ML).
Referenced localities. Cheik Hamid, El Kreig (Alfieri 1920), El Arish, Magdaba (Alfieri 1920 sub *H. variegatus* Fabricius), Wadi Isla, Cheikh Hamid, El Kreig (Alfieri 1976 sub *Curimosphena villosus* Haag-Rutenberg), the Sinai (Andres 1931b, Hart 1891, Peyerimhoff 1907 t. Hart).
General distribution. Saharo-Turano-Sindian chorotype. North Africa, Senegal up to Sudan, Ethiopia, eastern Mediterranean, Arabia (Kaszab 1981).
Comments. The data on *Imatismus variegatus* (Fabricius, 1781) for the Sinai are based on determination mistakes. According to Alfieri (1976), it appears that specimens identified as *I. variegatus* represent *I. villosus*.

VI.1.8.6 Genus *Mesostena* Eschscholtz, 1831

VI.1.8.6.1 Mesostena (Mesostena) angustata angustata (Fabricius, 1775) (see Plate X.2)
Pimelia angustata Fabricius, 1775: 253.
Mesostena oblonga Solier, 1835a: 401.
Mesostena laevicollis Solier, 1835a: 402.
Mesostena punctipennis Solier, 1835a: 403.
Material examined. 7 km S Taba, 3.XII.1992, L.P. (1); 19 km S Taba, 3.XII.1992, L.P. (4); Nuweiba (dune), 28.XI.1992, L.P. (1); Dahab, 9.IV.1993, P. (2); 1 km W Dahab, 29.XI.1992, L.P. (4); 55 km NE Santa Katarina, 850 m, 2.XII.1992, L.P. (13); 35 km N Sharm el Sheikh, 30.XI.1992, L.P. (24); 15 km N Sharm el Sheikh, 30.XI.1992, L.P. (28); 7 km N Sharm el Sheikh, 29.XI.1992, L.P. (2); Sharm el Sheikh, 29.-30.XI.1992, L.P. (4); Dahab, 9.IV.1993, P. (1); Dahab, 20.IV.1995, P. (3), 22.IV.1995, P. (3) (ML); area around the oasis of Ain Khudra, 25.-29.XII.1997, 600-800 m, leg. W. Schawaller (6); Sharm el Sheik, around Shark's Bay, 1.-4.I.1998, leg. W. Schawaller (4) (SMNS).
Referenced localities. All the peninsula (Alfieri 1976), Ga'a desert, Wadi Gergir, Wadi Budr (Andres 1920 sub *Mesostena laevicollis*), Ain Musa, Wadi Werdan, Wadi Gharandel, Ga'a desert, Tor, Wadi Gergir, Wadi Budr, Wadi Chaschibi (Andres 1920), between Suez and Abu Selima (Bodenheimer & Theodor 1929), El Kubri, El Qantara (Boyd 1917), the Sinai (Andres 1931b, Crotch 1872 sub *M. oblonga*, Gridelli 1933b, 1937, 1938, 1940, 1950, Hart 1891 sub *Mesotena punctipennis*, Holdhaus 1919 sub *M. laevicollis*, Koch 1934, 1940c, Peyerimhoff 1948a), Tor (Koch 1935a),

Wadi eth-Thal, Wadi Gharandel (Peyerimhoff 1907), Wadi el-Ain, Reidan Esqua'ah, Nouheibeh, Wadi Tabah, north of the Gulf of Aqaba (Peyerimhoff 1907 sub *M. laevicollis*), St Katherine Protectorate (Zalat et al. 2008); St Katharine, Sharm el Sheikh, Taba Protectorate (Hassan et al. 2017).

General distribution. *Mesosetena angustata angustata* (Fabricius 1775): Saharo-Sahelo-Arabian chorotype: Sahara, Sudan, Nigeria, Eritrea, the Sinai, eremic Palestine (Koch 1940c), Syria, southern Jordan (Schawaller 1982). – *Mesostena angustata deserticola* Soldati 2009: Qatar.

Comments. Koch (1940c) recognises several subspecies in the wide range, among which Peyerimhoff (1948a) counts *M. longicollis* Lucas, 1858 as a nominal subspecies. Furthermore *M. angustata deserticola* Soldati, 2009 is described from Qatar. This species is significantly ($P = 5.10^{-4}$) more prevalent at the South Sinai than at the North Sinai (Table II.1, Map VI.5).

VI.1.8.6.2   Mesostena (Mesostena) puncticollis Solier, 1835 (see Plate X.3)
Mesostena puncticollis Solier, 1835a: 405.

Material examined. Dahab, 20.IV.1995, P. (6); Dahab, 22.IV.1995, P. (5) (TP, ML).

Referenced localities. Wadi Isla (Alfieri 1976), south of Ain Musa (Andres 1920), the Sinai (Andres 1931b, Gridelli 1931 t. Andres).

General distribution. NE-African-Sindian chorotype: Turkmenistan, Iran, Iraq, Syria, Jordan, whole Arabian Peninsula (Koch 1940c, Kaszab, 1981), Israel (Krasnov & Ayal 1995), Azerbaijan, Turkey (Iwan et al. 2020), Sudan (Ardoin 1972a), Somalia (Koch 1940c), Djibouti (ML).

VI.1.8.6.3   Mesostena (Mesostenopa) peyerimhoffi Gridelli, 1938
Mesostenopa gracilis Peyerimhoff, 1907: 31 nec *M. gracilis* Gredler, 1878: 514.
Mesostena (Mesostenopa) peyerimhoffi Gridelli, 1938: 109, replacement name.

Material examined. SE Sinai, Wadi Umm Ahmed, 8.-13.XII.1991, Fichtner & Leidenroth (1) (SMNS).

Referenced localities. The Sinai (Alfieri 1976, Andres 1931b sub *Mesostenopa gracilis*), Wadi Hebran (Andres 1920 sub *Mesostenopa gracilis*), Ain al-Houdra (Gridelli 1938, Koch 1935a sub *Mesostenopa gracilis*), Ain el-Houdra ("Hazeroth") (Innes Bey 1909 t. Peyerimhoff, Peyerimhoff 1907 sub *Mesostenopa gracilis*), eastern Sinai: Ain el Houdra (Schatzmayr & Koch 1934 sub *Mesostenopa gracilis*).

General distribution. Endemic to the eastern Sinai (Schatzmayr & Koch 1934).

VI.1.8.6.4   Mesostena (Mesostenopa) picea sinaitica (Schatzmayr & Koch, 1934) (see Plate X.4)
Mesostenopa habessinica sinaitica Schatzmayr & Koch, 1934: 17.

Material examined. 10 km S Taba, 27.XI.1992, L.P. (2); 13 km S Taba, 3.XII.1992, L.P. (1); 3 km E Ain el Furtaga, 28.XI.1992, L.P. (1); 20 km S Nuweiba, 27.XI.1992, L.P. (1); 12 km NW Dahab, 30.XI.1992, L.P. (1); Dahab, 9.IV.1993, P. (2); 18 km S Dahab, 29.XI.1992, L.P. (2); 7 km NE Santa Katarina, 2.XII.1992, L.P. (1); 35 km N Sharm el Sheikh, 30.XI.1992, L.P. (1); Santa Katarina, 16.IV.1995, P. (1); Santa Katarina, Plain of Raha, 17.IV.1995, P. (1); Sharm el Sheik, 1.-4.I.1998,

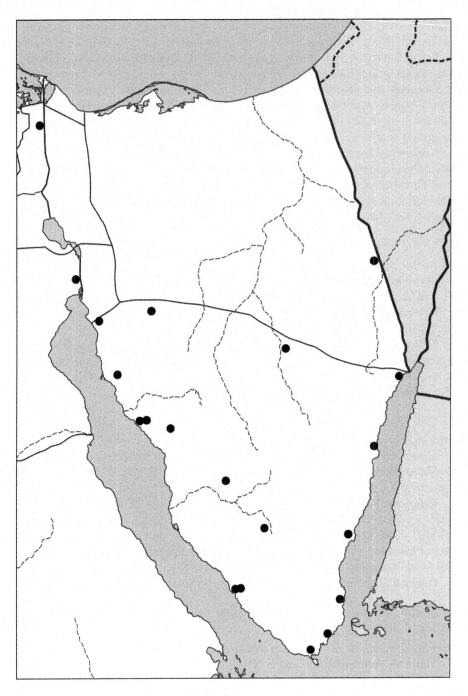

MAP VI.5 Distribution map of *Mesostena angustata angustata* in the Sinai Peninsula.

leg. W. Schawaller (1) (ML); Santa Katarina, 29.IV.1987, Monastery of St. Catherine, remaining ponds, leg. M. Balke (1) (SMNS).
Referenced localities. Wadi Karam (Alfieri 1976), Wadi Isla, Wadi Hebran (Koch 1935a), Wadi Isla, between Tor and Santa Katarina (Schatzmayr & Koch 1934 sub *Mesostenopa habessinica sinaitica*), the Sinai (Koch 1934 sub *Mesostenopa habessinica sinaitica*).
General distribution. Endemic to the Sinai (Koch 1940c).
Comments. According to Koch (1940c), this species is divided into seven subspecies distributed over a vast part of the Sahara and of the Arabian Peninsula. *M. picea sinaitica* was known up to now from the south-west of the Sinai (Koch 1940c). Koch (1935a) mentioned it also from Gebel Elba, but corrected this information later (Koch 1940c). The information given by Alfieri (1976) according to which this subspecies occurs also in Lower Egypt seems to be questionable.

VI.1.8.7 Genus *Micipsa* Lucas, 1855

VI.1.8.7.1 Micipsa (Micipsa) burtoni Baudi di Selve, 1881 (see Plate XI.1)
Micipsa burtoni Baudi di Selve, 1881: 277.
Material examined. El Arish, 26.12.1956, T. Ephter (1) (ML),
Referenced localities. "Sinai: Kid" (Koch 1935a). St Katherine Protectorate, V.-VI. 2010 (Norfolk et al. 2012).
General distribution. Levantine chorotype: Syria (Damascus) (Baudi di Selve 1881), Israel (Negev) (Ayal & Merkl 1994), Sinai (Koch 1935a).

VI.1.8.7.2 Micipsa (Micipsa) douei douei Lucas, 1856 (see Plate XI.2)
Micipsa douei Lucas, 1856: xlvi.
Micipsa grandis Kraatz, 1865: 108.
Material examined. Around the oasis of Ain Khudra, 25.-29.XII.1997, 600-800 m, leg. W. Schawaller (1); Sharm el Sheik around Shark's Bay, 1.-4.I.1998, leg. W. Schawaller (1) (SMNS).
Referenced localities. Wadi Isla (Alfieri 1976), Wadi Gergir (Andres 1920), the Sinai (Gebien 1906, Gridelli 1930), Tor (Koch 1935a), Wadi Chamileh (Heyden 1899), Wadi Khamileh (Peyerimhoff 1907 t. Heyden).
General distribution. SW-Asiatic chorotype: Lebanon (Beirut) (Kraatz 1865), the Sinai, Egypt (Andres 1920), Libya (Cyrenaica) (Gridelli 1930), Algeria, Jordan (Iwan et al. 2020), Sudan (Gebien 1906). *M. douei distincticornis* Peyerimhoff, 1933: Algeria (Peyerimhoff, 1933), Libya (Peyerimhoff, 1948a).

VI.1.8.7.3 Micipsa (Micipsa) cf. *philistina* Reiche & Saulcy, 1857 (see Plate XI.3)
Micipsa philistina Reiche & Saulcy, 1857: 212.
Material examined. Santa Katarina, 1500 m, 1.-2.XII.1992, L.P. (4); 1 km E Santa Katarina Monastery, 1800 m, 1.XII.1992, L.P. (5); 7 km NE Santa Katarina, 1400 m, 2.XII.1992, L.P. (1) (ML); Santa Katarina, 16.4.1995, P. (2); Santa Katarina, Plain of Raha, 17.4.1995, P. (3) (ML, TP).
Referenced localities. Kid (Andres 1931b, Koch 1935a), the Sinai (Crotch 1872).
General distribution. SW-Asiatic chorotype: The Sinai (Koch 1935a), Jordan, Iraq, "Syria" (Kaszab 1981), Israel (Krasnov & Ayal 1995).

Comments. The specimens examined are clearly different from those from Israel. Type locality is Nablus, West Bank. Maybe they are two different species, but this can only be clarified by a genus revision.

VI.1.8.7.4   Micipsa (Micipsa) schaumii Kraatz, 1865 (see Plate XI.4)
Micipsa schaumii Kraatz, 1865: 111.
   Referenced localities. The Sinai (Crotch 1872), Hadjar er-Reqab (Peyerimhoff 1907).
   General distribution. Southern Levantine chorotype: Lower Egypt (Alfieri 1976), the Sinai (Peyerimhoff 1907).

VI.1.8.8   Genus *Oterophloeus* Desbrochers des Loges, 1881
VI.1.8.8.1   Oterophloeus alveatus peyerimhoffi Koch, 1935 (see Plate XII.1)
Oterophloeus alveatus peyerimhoffi Koch, 1935a: 23, 24.
   Material examined. 7 km NE Santa Katarina, 2.XII.1992, L.P. (1) (ML).
   Referenced localities. Wadi Isla (Alfieri 1976).
   General distribution. *O. alveatus alveatus* Peyerimhoff, 1931: 95: Hoggar (Peyerimhoff 1931), south Morocco, former Spanish Sahara, Mauritania (Español 1952, Gridelli 1952, Kocher 1958, 1964, Reymond 1952), Libya (Iwan et al. 2020). – *O. alveatus peyerimhoffi*: Southern Levantine chorotype: Lower Egypt (Koch 1935a), the Sinai (Alfieri 1976).

VI.1.8.8.2   Oterophloeus haagii orientalis Koch, 1935 (see Plate XII.2)
Oterophloeus haagi orientalis Koch, 1935a: 25.
   Referenced localities. Wadi Isla (Alfieri 1976), south of Ain Musa up to Wadi Werdan (Andres 1920 sub *Tentyrina haagi*), El Arish (Koch 1935a), Hadjar er-Reqab (Peyerimhoff 1907 sub *Tentyriina haagi*).
   General distribution. Southern Levantine chorotype: From Cairo up to the Sinai (Alfieri 1976). *O. h. haagii* (Kraatz, 1865) in Lower Egypt (Koch 1935a).

VI.1.8.9   Genus *Oxycara* Solier, 1835
VI.1.8.9.1   Oxycara (Oxycara) ardoini Kaszab, 1979 (see plate XII.3)
Oxycara (Oxycara) ardoini Kaszab, 1979: 285.
   Material examined. Area around Ain Khudra oasis, 25.-29.XII.1997, 600-800 m, leg. W. Schawaller (7); 30 km W Nuweiba, Gebel Barqa, 4.-6.II.1997 600 m, leg. Iglesias & Leidenroth (3) (SMNS).
   Referenced localities. Reported from the Sinai for the first time.
   General distribution. Southern Levantine chorotype: Southern Jordan (Kaszab 1979), the Sinai, Israel (the Negev) (SMNS).

VI.1.8.9.2   Oxycara (Oxycara) productum Peyerimhoff, 1907
Oxycara productum Peyerimhoff, 1907: 33.
   Referenced localities. Wadi Mitla, Wadi Isla (Alfieri 1976), the Sinai (Andres 1931b t. Peyerimhoff), Ain el-Houdra (Innes Bey 1909 t. Peyerimhoff, Peyerimhoff 1907).
   General distribution. Endemic to the Sinai (Alfieri 1976).

VI.1.8.9.3 Oxycara (Oxycara) pygmaeum (Reiche & Saulcy, 1857) (see Plate XII.4)

Melancrus pygmaeus Reiche & Saulcy 1857: 194.

Referenced localities. Wadi Isla (Alfieri 1976), between Suez and Abu Selima (Bodenheimer & Theodor 1929), between Wadi Werdan and Wadi Feiran (Andres 1920), the Sinai (Gridelli 1930, 1933a t. Andres).

General distribution. North African chorotype: Libya (Gridelli 1933a) up to the Sinai (Alfieri 1976).

Comments. This species had been described from the Dead Sea and from "Syrie". However, it does not seem that it occurs there, too. Since the authors did not determine a holotype, a lectotype was designated by Lillig & Pavlíček (2003): Lectotype: Sex not studied: "M. morte" "Type" "Type" [red] "Muséum Paris, Coll. Reiche" "Muséum Paris, 1932, Coll. Bedel, Soc. Ent. De France" *"pygmaeum* Reiche" (MNHNP). Paralectotypes: "Syrie" "Type" "Type" [red] "Muséum Paris, Coll. Reiche" "Muséum Paris, 1932, Coll. Bedel, Soc. Ent. De France" (5) (all in MNHNP).

VI.1.8.9.4 Oxycara (Symphoxycara) peyerimhoffi Lillig, 2001 (see Plate XIII.1)

Oxycara peyerimhoffi Lillig, 2001: 369.

Material examined. 19 km S Taba, 3.XII.1992, L.P. (1); 20 km S Taba, 27.XI.1992, L.P. (1); 5 km N Nuweiba, 28.XI.1992, L.P. (4); 3 km N Nuweiba, 3.XII.1992, L.P. (1); Dahab, 9.IV.1993, P. (1); 1 km W Dahab, 29.XI.1992, L.P. (1); 15 km N Sharm el Sheikh, 30.XI.1992, L.P. (3); Regam in W. Kid, 25.X.<19>68, leg. Broza (3); Eim Furghaga, 1.VII.1970, leg. Broza (2) (TAU) (Lillig 2001).

Referenced localities. The collection sites given for *O. breviusculum* Fairmaire and *O. subcostatum* (Guérin-Méneville) should refer to this species: Wadi Isla, Wadi Um Mitla (Alfieri 1976 sub *O. subcostatum*), Wadi Gergir, Wadi Budr, Ta albi, Abu Cscheib, Chaschibi, Wadi Ab-Orta (Andres 1920 sub *O. subcostata*), southern part of the Sinai Peninsula (Andres 1931b sub *O. subcostatum*), Gala el Aqaba (Gridelli 1953 sub *O. breviusculum*), Tor (Koch 1935a sub *O. breviusculum*), Wadi el-Ain, Wadi Taba (Peyerimhoff 1907 sub *Oxycara subcostata* Guér.).

Comments. This species is related to *Oxycara breviusculum* Fairmaire, 1892 (Lillig 2001). *O. peyerimhoffi* is significantly ($P = 0.01$) more prevalent at the South Sinai than at the North Sinai (Table II.1, Map VI.6).

VI.1.8.10 Genus *Prochoma* Solier, 1835

VI.1.8.10.1 Prochoma (Prochoma) audouini Solier, 1835 (see Plate XIII.2)

Prochoma audouini Solier, 1835a: 395.

Material examined. 2 km N Nuweiba (coastal oasis), in spider's web, 28.XI.1992, L.P. (1) (ML).

Carl (1994) already published these data.

General distribution. NE-African-Sindian chorotype: Iraq, Iran, Syria, Jordan, the Sinai (Carl 1994), Israel (Jerusalem) (Koch 1940b).

Comments. Nuweiba is the most western located collection site of the genus.

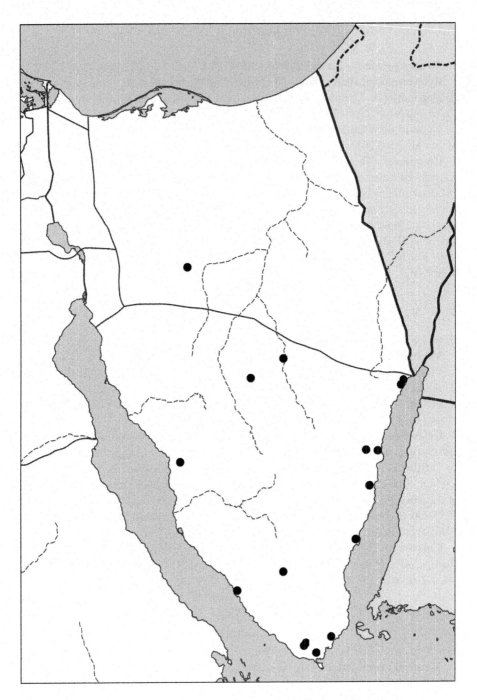

MAP VI.6 Distribution map of *Oxycara peyerimhoffi* in the Sinai Peninsula.

VI.1.8.11 Genus *Scelosodis* Solier, 1835

VI.1.8.11.1 Scelosodis castaneus castaneus (Eschscholtz, 1831) (see Plate XIII.3)
Cratopus castaneus Eschscholtz, 1831: 8.
Material examined. Dahab, 9.IV.1993, P. (1); Dahab, 20.IV.1995, P. (8) (ML, TP); area around oasis Ain Khudra, 25.-29.XII.1997, 600-800 m, leg. W. Schawaller (11) (9 SMNS, 2 ML).
Referenced localities. Wadi Isla (Alfieri 1976), Ain Musa, Wadi Werdan, Wadi Gharandel, Wadi Schellal, Wadi Mokatteb, Wadi Feiran (Andres 1920), the Sinai (Andres 1931b, Girard & Pierre 1965, Gridelli 1950, Koch 1934, Peyerimhoff 1933, 1936), Katia (Boyd 1917), Wadi Isla, Tor (Koch 1935a), Ras abou Zenimeh, Wadi Gneh, Ain el-Houdra (Peyerimhoff 1907).
General distribution. *S. castaneus castaneus*. Saharian chorotype: Upper and Lower Egypt, Libya (Cyrenaica), the Sinai (Girard & Pierre 1965), Algeria (Peyerimhoff 1933).
S. castaneus soudanicus Girard & Pierre, 1965: 135: Sahel from Khartoum up to Air (Ardoin 1972a, Español 1973, Girard & Pierre 1965).

VI.1.8.12 Genus *Schweinfurthia* Andres, 1922

VI.1.8.12.1 Schweinfurthia alfierii Schatzmayr & Koch, 1934
Schweinfurthia alfierii Schatzmayr & Koch, 1934: 20.
Referenced localities. Wadi Meghara (Alfieri 1976), Tor (Schatzmayr & Koch 1934).
General distribution. Endemic to the Sinai (Alfieri 1976).

VI.1.8.12.2 Schweinfurthia sinaitica Andres, 1922
Schweinfurthia sinaitica Andres, 1922: 26.
Referenced localities. Ain Musa (Andres 1920 sub n. gen., n. sp., Andres 1922, 1931b), the Sinai (Alfieri 1976, Schatzmayr & Koch 1934).
General distribution. Endemic to the Sinai. Only the holotype is known until now. No other specimens have been collected yet.

VI.1.8.13 Genus *Tentyria* Latreille, 1802

VI.1.8.13.1 Tentyria discicollis Reiche & Saulcy, 1857 (see Plate XIII.4)
Tentyria discicollis Reiche & Saulcy, 1857: 207.
Referenced localities. The Sinai (Crotch 1872).
General distribution. Levantine chorotype: Syria, Jordan (Kaszab 1981, Katbeh-Bader 1996, Schawaller 1982), Israel (Ayal & Merkl 1994), the Sinai (Crotch 1872).
Comments. This species is not rare in the northern Negev. Consult Schawaller (1982) to differentiate between *T. discicollis* and *T. laticollis* Kraatz, 1865.

VI.1.8.13.2 Tentyria punctatostriata Solier, 1835
Tentyria punctatostriata Solier, 1835a: 337.
Referenced localities. Mohammedia (Boyd 1917).
General distribution. North African chorotype: Eastern Cyrenaica (Gridelli 1930), Lower Egypt (Koch 1935a), northern Sinai (Boyd 1917).

VI.1.8.13.3   Tentyria sinaitica Peyerimhoff, 1907 (see Plate XIV.1)
Tentyria sinaitica Peyerimhoff, 1907: 32.
Material examined. 27 km NE Santa Katarina, 1300 m, 2.XII.1992, L.P. (1); Mt. Moses (Gebel Musa), 7.-8.IV.1993, P. (1); Santa Katarina, 16.IV.1995, P. (5); Santa Katarina, Wadi Tlach, 16.IV.1995, P. (8); Santa Katarina, Plain of Raha, 17.IV.1995, P. (1) (ML, TP).
Referenced localities. Cheik Hamid, Wadi Meghara (Alfieri 1976), Djebel Serbal (Andres 1931b t. Peyerimhoff), Abu Hamaida, Mear (Andres 1931b), Djebel Serbâl in the Sinai mountains (Innes Bey 1909 t. Peyerimhoff, Peyerimhoff 1907), the Sinai (Koch 1935a), St Katherine Protectorate (Zalat et al. 2008).
General distribution. Endemic to the South Sinai.

VI.1.8.14   Genus *Tentyrina* Reitter, 1900
VI.1.8.14.1   Tentyrina orbiculata subsulcata (Reiche & Saulcy, 1857) (see Plate XIV.2)
Tentyria subsulcata Reiche & Saulcy, 1857: 203.
Tentyria aegyptiaca Solier, 1835a: 334.
Material examined. Nuweiba, III.1997, leg. N. Rech (1); Wadi Gharandel, IV.1993, leg. Ullrich (1) (ML).
Referenced localities. Wadi Mezara, Um Shousha (Alfieri 1920 sub *Tentyria aegyptiaca*), Wadi Meghara, Wadi Isla (Alfieri 1976), the Sinai (Andres 1931b sub *T. orbiculata* (Fabricius), Grimm 1991), El Arish, Romani (Andres 1931b), between Suez and Abu Selima (Bodenheimer & Theodor 1929), Mohammedia, Romani, Mazar (Boyd 1917 sub *T. aegyptiaca*), El Arish, Romani (Gridelli 1929b), Wadi Feiran (Koch 1935a sub *Tentyrina boehmi subsulcata*), Wadi Amara (Peyerimhoff 1907 t. Heyden sub *T. orbiculata* (Fabricius).
General distribution. *T. orbiculata orbiculata* (Fabricius, 1775: 253): Lower Egypt (Koch 1940c). – *T. o. subsulcata*. East Mediterranean chorotype: Egypt, the Sinai, Israel, Lebanon, Cyprus (Grimm 1991) and Syria (Iwan et al. 2020).

VI.1.8.14.2   Tentyrina palmeri palmeri (Crotch, 1872) (see Plate XIV.3)
Tentyria palmeri Crotch, 1872: 267.
Tentyria glabra auct. nec (Fabricius, 1775: 253).
Tentyria giraffa Allard, 1883: 22.
Material examined. 19 km S Taba, 3.XII.1992, L.P. (1); 10 km N Nuweiba, 28.XI.1992, L.P. (1); 40 km NNW Dahab, 2.XII.1992, L.P. (10); 10 km NW Dahab, 2 XII.1992, L.P. (2); 55 km NE Santa Katarina, 2.XII.1992, L.P. (2); Nuweiba, 27.III.1996, leg. N. Rech (2); Nuweiba, III.1997, leg. N. Rech (2) (ML); Sharm el Sheik around Shark's Bay, 1.-4.I.1998, leg. W. Schawaller (2); zone between Dahab and Nuweiba, 16.-21.11.1995, 500-800 m, leg. S. & C. Leidenroth (2); 30 km NW Nuweiba, Gebel Barqa, 4.-6.II.1997, 600 m, leg. Iglesias & Leidenroth (5); Gebel el Gunna, Abu Shauarib, 8.II.1997, 700 m, leg. C. & S. Leidenroth (3) (SMNS).
Referenced localities. Wadi Karam (Alfieri 1976 sub *Tentyria palmeri*), the Sinai (Andres 1931b, Blair 1931, Crotch 1872, Gebien & Borchmann 1927, Kaszab 1981, Pierre 1961b, Peyerimhoff 1948a, Schuster 1938, Schuster & Gebien 1938), Wadi Feiran (Heyden 1899 sub *Tentyria glabra*, Peyerimhoff 1907 t. Heyden), Wadi el-Ain, Nuweiba, northern Gulf of Aqaba (Peyerimhoff 1907).

General distribution. *T. palmeri palmeri*: NE-African-Sindian chorotype: Eastern Sahara from Libya to Egypt, the Sinai, Saudi Arabia, Iraq, Iran (Kaszab 1981, Koch 1940c), Syria and Jordan (Schawaller 1982). – *T. palmeri thomasi* (Blair, 1931: 22): Oman, Qatar, Saudi Arabia (Soldati 2009).

VI.1.8.15   Genus *Thraustocolus* Kraatz, 1865

VI.1.8.15.1   Thraustocolus (Thraustocolus) leptoderus (Kraatz, 1865)
(see Plate XIV.4)

Calobamon leptoderus Kraatz, 1865: 106.
Material examined. Sharm el Sheik around Shark's Bay, 1.-4.I.1998, leg. W. Schawaller (1) (SMNS).
Referenced localities. Wadi Gueh (Andres 1931b, Peyerimhoff 1907), the Sinai (Kaszab 1979, Schuster 1934).
General distribution. Southern Levantine chorotype: Egypt, the Sinai (Alfieri 1976), Dead Sea (Kaszab 1979).

VI.1.8.16   Genus *Trichosphaena* Reitter, 1916

VI.1.8.16.1   Trichosphaena perraudierei perraudierei (Marseul, 1876)
(see Plate XV.1)

Himatismus perraudierei Marseul, 1867: xxxix.
Referenced localities. Wadi Karam (Alfieri 1976), El Arish, Wadi Um Helal (Koch 1935a).
General distribution. *T. perraudierei perraudierei*: Saharo-Sahelo-Arabian chorotype: Sahara up to Sudan, Saudi Arabia (Kaszab 1981), the Negev (Bytinski-Salz 1955a). – *T. perraudieri peyerimhoffi* Gridelli, 1950: Mauritania (Adrar), Niger (Air), Chad (Ennedi, Mortcha) (Español 1973, Gridelli 1950, 1952).

VI.1.9   Tribe Zophosini Solier, 1834

VI.1.9.1   Genus *Zophosis* Latreille, 1802

VI.1.9.1.1   Zophosis (Hologenosis) nigroaenea Deyrolle, 1867 (see Plate XV.2)
Zophosis nigroaenea Deyrolle, 1867: 181.
Zophosis aelanitica Peyerimhoff, 1907: 26.
Material examined. 1 km W Dahab, 29.XI.1992, L.P. (1) (ML).
Referenced localities. Prope Aelanam Judaeorum (Innes Bey 1909 t. Peyerimhoff, Peyerimhoff 1907 sub *Z. aelanitica*), Gulf of Aqaba (Koch 1935a sub *Z. aelanitica*).
General distribution. North African chorotype: Mauritania up to the Gulf of Aqaba (Penrith 1982b).
Comments. The species have been described as *Z. aelanitica* by Peyerimhoff (1907) after a single specimen from the Sinai part of the Gulf of Aqaba ("in arena mobile prope Aelanam Judaeorum").

VI.1.9.1.2   Zophosis (Oculosis) punctata punctata Brullé, 1832 (see Plate XV.3)
Zophosis punctata Brullé, 1832: 191.
Zophosis maillei Solier, 1834: 613.

Referenced localities. The Sinai (Andres 1931b, Crotch 1872 sub *Z. maillei*), Tor (Penrith 1983).

General distribution. *Z. punctata punctata*: Centralasiatic-Mediterranean chorotype: North Africa from Morocco to Libya, in Europe from Spain to Greece, in Asia from Turkey to Pakistan (Penrith 1983), Uzbekistan, China (Xinjiang) (Iwan et al. 2020) and Saudi Arabia (Kaszab 1981). – *Z. p. pygmaea* Solier, 1834: Lower Egypt, the Negev, Kuwait (Penrith 1983). – *Z. p. alborana* Baudi di Selve, 1883: Alborana island facing the Spanish southern coast (Penrith 1983).

Comments. Transition forms between *Z. p. punctata* and *Z. p. pygmaea* have been observed in Israel, the Sinai and Iraq (Penrith 1983).

VI.1.9.1.3 Zophosis (Septentriophosis) bicarinata cf. *ghilianii* Deyrolle, 1867 (see Plate XV.4)

Zophosis ghilianii Deyrolle, 1867: 196.

Material examined. Nuweiba (dune), 28.XI.1992, L.P. (4); 1 km W Dahab, 29.XI.1992, L.P. (4); 55 km NE Santa Catherine, 850 m., 2.XII.1992, L.P. (1); 15 km N Sharm el Sheikh, 30.XI.1992, L.P. (3); 7 km N Sharm el Sheikh, 29.XI.1992, L.P. (7); Dahab, 20.4.1995, P. (2) (ML, TP); Sharm el Sheik around Shark's Bay, 1.-4.I.1998, leg. W. Schawaller (4 dead) (SMNS).

General distribution. *Z. bicarinata* Solier s. l., composed of ten subspecies, is distributed from the Canary Islands to the Arabian Peninsula. Unfortunately, the type locality of *Z. b. ghilianii* is unknown (Penrith 1986).

VI.1.9.1.4 Zophosis (Septentriophosis) complanata Solier, 1834 (see Plate XVI.1)

Zophosis complanata Solier, 1834: 626.

Material examined. East of Suez City, V.<19>71, leg. D. Simon (1) (TAU).

Referenced localities. Wadi Mezara, Um Shousha (Alfieri 1920), south of Ain Musa up to Wadi Werdan (Andres 1920), between Suez and Abu Selima (Bodenheimer & Theodor 1929), Pharaoh's Baths (= Hammam Fir'aoun), Wadi Gorondel, Ga'a plain (Penrith 1984).

General distribution. SW-Asiatic chorotype: Egypt, the Sinai, Jordan, Gaza, Saudi Arabia, Iraq, Iran (Penrith 1984), Israel (Bytinski-Salz 1955a).

VI.1.9.1.5 Zophosis (Septentriophosis) farinosa Blair, 1914

Zophosis farinosa Blair, 1914: 483, replacement name.

Zophosis testudinaria Solier, 1834: 622 nec *Erodius testudinarius* Fabricius, 1787: 215.

Referenced localities. South of Ain Musa (Andres 1920), the Sinai (Andres 1931b sub *Z. testudinaria* Fabricius, Kaszab 1981), Abou Zenneh (between Suez and Feiran) (Peyerimhoff 1907 sub *Z. testudinaria* Fabricius).

General distribution. NE-African-Sindian chorotype: NE Saudi Arabia, Jordan, Iraq, Iran (Penrith 1984), the Sinai (Kaszab 1981).

VI.1.9.1.6 Zophosis (Septentriophosis) lethierryi Deyrolle, 1867 (see Plate XVI.2)

Zophosis lethierryi Deyrolle, 1867: 211.

Referenced localities. El Arish, Bugnia (= Umm Bugma) (Penrith 1982a).

General distribution. North African chorotype: Northern Algeria, Tunisia, Libya (Tripolitania), the Sinai (Penrith 1982a).

VI.1.9.1.7 Zophosis (Septentriophosis) personata Erichson, 1841 (see Plate XVI.3)
Zophosis personata Erichson, 1841: 176.
 Zophosis depressipennis Lucas, 1858: ccxxii.
 Referenced localities. Mohammedia, Mazar (Boyd 1917 sub *Z. depressipennis*), Umm Bugma (Penrith 1982a).
 General distribution. Saharian chorotype: Northern Sahara, the Sinai (Penrith 1982a).

VI.1.9.1.8 Zophosis (Septentriophosis) pharaonis pharaonis Reitter, 1916 (see Plate XVI.4)
Zophosis pharaonis Reitter, 1916c: 90, replacement name.
 Zophosis rotundata Deyrolle, 1867: 215 nec Ménétriés, 1849: 218.
 Material examined. Romano, 20.VI.1970, leg. Bytinski-Salz (1); Biz Gif, 24.IV.1968, leg. H. Schweiger (1) (TAU).
 Referenced localities. South of Ain Musa, Wadi Werdan, between Wadi Werdan and Hawara, Wadi Gharandel (Andres 1920), the Sinai (Andres 1921, 1931b, Blair 1934 sub *Z. rotundata* Ménétriés, Gridelli 1933a, Schatzmayr 1938), Wadi Meghara (Penrith 1982a), Pharaoh's Baths (= Hammam Fir'aoun) (Walker 1871, determination uncertain).
 General distribution. *Z. pharaonis pharaonis*: SW-Asiatic chorotype: Lower Egypt, the Sinai, NW Saudi-Arabia (Penrith 1982a), Israel (Ayal & Merkl 1994). – *Z. pharaonis simplex* Kaszab, 1981: 319, Southern Arabia (Penrith 1984).

VI.1.9.1.9 Zophosis (Septentriophosis) plana (Fabricius, 1775) (see Plate XVII.1)
Erodius planus Fabricius, 1775: 259.
 Zophosis carinata Solier, 1834: 630.
 Material examined. Santa Katarina, 18.V.1970, leg. Bytinski-Salz (1) (TAU).
 Referenced localities. Wadi Hebran, Ga'a up to Tor, Wadi Chaschibi (Andres 1920 sub *Z. carinata*), the Sinai (Andres 1931b sub *Z. carinata*, Boehm 1909, Gridelli 1952 sub *Z. carinata*, Koch 1934 sub *Z. carinata*, Peyerimhoff 1934, 1936), El Arish (Boyd 1917), Wadi Feiran (Ardoin 1972a, Heyden 1899 sub *Z. carinata*), Nachlé (Heyden 1899 sub *Z. carinata*), Qala'at en-Nakhel (Peyerimhoff 1907 t. Heyden sub *Z. carinata*), Tor (Koch 1935a sub *Z. carinata*), En Nkeyaz, Nachle (Penrith 1982a).
 General distribution. Saharian chorotype: Central, northern and eastern Sahara, the Sinai (Penrith 1982a).

VI.2   Subfamily Blaptinae Leach, 1815

VI.2.1   Tribe Blaptini Leach, 1815

VI.2.1.1   Genus *Blaps* Fabricius, 1775

VI.2.1.1.1   Blaps bifurcata Solier, 1848 (see Plate XVII.2)
Blaps bifurcata bifurcata Solier, 1848: 348.
 Blaps bifurcata andresi Koch, 1935a: 75.
 Referenced localities. Gebel Katherin (Alfieri 1976), Wadi Helal (Alfieri 1976 sub *B. bifurcata andresi*), El Arish (Andres 1931b sub *B.* spec. pr. *bifurcata*, Gridelli 1952 t. Koch, Koch 1935a sub *B. bifurcata andresi*), Romani (Boyd 1917).

General distribution. *B. bifurcata bifurcata*: North African chorotype: Israel (Negev) (Ayal & Merkl 1994), Libya (Fezzan), Egypt, the Sinai (Gridelli 1952). – *B. bifurcata mirei* Gridelli, 1952: 92: Mauritania up to Chad (Rougon & Ardoin 1976), Sudan (HJB, ML). – *B. bifurcata strauchi* Reiche, 1861: 88 (= *B. bifurcata gridellii* Pierre, 1961: 1042; Condamine et al. 2011): Algeria, Tunisia, Libya, Chad (Tibesti), Morocco, former Spanish Sahara (Gridelli 1952) and Niger (Ténéré) (Pierre 1961b).

VI.2.1.1.2 Blaps judaeorum Miller, 1861
Blaps judaeorum Miller, 1861: 178
   Blaps pharao Seidlitz, 1893: 266.
   Referenced localities. Northern Sinai (Andres 1931b sub *B. pharao*), El Arish (Koch 1935a sub *B. pharao*).
   General distribution. SW-Asiatic chorotype: Egypt, Iraq, Israel, Lebanon, Syria, the Sinai (Iwan et al. 2020).
   Comments. The synonymy of *B. judaeorum* and *B. pharao* was established by Condamine et al. (2011).

VI.2.1.1.3 Blaps nitens laportei Ardoin, 1973 (see Plate XVII.3)
Blaps nitens laportei Ardoin, 1973: 291, replacement name.
   Blaps sulcata Laporte de Castelnau, 1840: 200 nec Fabricius, 1775: 254.
   Referenced localities. Magdaba (Alfieri 1920, Gridelli 1930 t. Alfieri), the Sinai (Crotch 1872 sub *B. sulcata*), Wadi Bel Maï, Wadi El Arish (Heyden 1899 sub *Blaps sulcata*).
   General distribution. North African chorotype: Libya (Cyrenaica), Egypt (Koch 1944), Israel (Negev) (Ayal & Merkl 1994), the Sinai.
   Comments. Ardoin (1973) discusses the nomenclature of the superspecies complex.

VI.2.1.1.4 Blaps polychresta (Forskål, 1775) (see Plate XVII.4)
Tenebrio polychrestus Forskål, 1775: 75.
   Blaps sulcata Fabricius, 1775: 154 nec Solier, 1848: 344.
   Referenced localities. Wadi Ain El Gederat (Alfieri 1920, 1976), Wadi Isla (Alfieri 1976), oasis of Ain Musa (Andres 1920), Ain Musa (Peyerimhoff 1907 t. Walker sub *Blaps sulcata* Fabricius, Walker 1871 sub *Blaps sulcata* Fabricius), Wadi El Arish (Peyerimhoff 1907 t. Heyden sub *Blaps sulcata* Fabricius).
   General distribution. North African chorotype: Libya (Cyrenaica), Egypt (Koch 1937), the Negev (Bytinski-Salz 1955a), the Sinai (Alfieri 1976).
   Comments. Confusions are possible due to the name jumble (cf. *Blaps nitens laportei* Ardoin, 1973).

VI.2.1.1.5 Blaps wiedemanni Solier, 1848
Blaps wiedemanni Solier, 1848: 350.
   Referenced localities. Wadi Bel Maï (Heyden 1899).
   General distribution. SW-Asiatic chorotype: Libya, Egypt, Saudi Arabia, Yemen (Kaszab 1982), Israel (Krasnov & Ayal 1995), the Sinai (Heyden 1899).

VI.2.2 Tribe Dendarini Mulsant & Rey, 1854

VI.2.2.1 Genus *Dendarus* Dejean, 1821

VI.2.2.1.1 Dendarus (Rhizalemus) calcaroides (Peyerimhoff, 1907) (see Plate XVIII.1)

Phylax (Rhizalemus) calcaroides Peyerimhoff, 1907: 36.
Material examined. Mt. Moses (Gebel Musa), 7.-8.IV.1993, P. (8) (ML, TP).
Referenced localities. Wadi Helal, Wadi Isla (Alfieri 1976), Monastery of St. Catherine (Andres 1931b t. Peyerimhoff), Abu Hamaida (Andres 1931b), the Sinai (Koch 1935a), and Mount Sinai, Djebel Katharina (Innes Bey 1909 t. Peyerimhoff, Peyerimhoff 1907).
General distribution. Endemic to South Sinai (Alfieri 1976).

VI.2.2.1.2 Dendarus (Rhizalus) syriacus (Reiche & Saulcy, 1857) (see Plate XVIII.2)

Bioplanes syriacus Reiche & Saulcy, 1857: 257.
Material examined. Santa Katarina, 1500 m, 1.-2.XII.1992, L.P. (1) (ML, TP); St. Katerine, 5.X.1992 (2) (ML, CGW).
Referenced localities. Wadi Helal, Wadi Isla (Alfieri 1976), the Sinai (Crotch 1872 sub *Bioplanes syriacus*).
General distribution. Levantine chorotype: Palestine,? Syria (Koch 1935b), Jordan (Peyerimhoff 1907), the Sinai, Egypt (Alfieri 1976).

VI.2.3 Tribe Opatrini Brullé, 1832

VI.2.3.1 Genus *Ammobius* Guérin-Méneville, 1844

VI.2.3.1.1 Ammobius rufus (Lucas, 1846) (see Plate XVIII.3)

Trachyscelis rufus Lucas, 1846: 339.
Referenced localities. El Kontella, Wadi El Arish (Alfieri 1976).
General distribution. Turano-Mediterranean chorotype: Moroccan Atlantic coast, coasts of the Meditarrenean Sea and of the Black Sea (Grimm 1991).

VI.2.3.2 Genus *Clitobius* Mulsant & Rey, 1859

VI.2.3.2.1 Clitobius grimmi Lillig, 2021 (see Plate XVIII.4)

Clitobius grimmi Lillig, 2021: 479.
Material examined. 14 km S Taba, 27.XI.1992, L.P. (1); 10 km N Nuweiba, 28.XI.1992, L.P. (3); Dahab, 28.XI.1992, L.P. (2), Dahab, 20.IV.1995, P. (2) (ML, TP); Sharm el Sheik around Shark's Bay, 1.-4.I.1998, leg. W. Schawaller (10) (SMNS); Golfe d'Aqabah, 18.2.02, 1 ex. (MNHN).
Referenced localities. Gulf of Aqaba (Peyerimhoff 1907 sub *C. oblongiusculus* var. *lineicollis*).
General distribution. Red Sea seashore chorotype: The Sinai and Hurghada, endemic to Egytian coast of the Red Sea.
Comments. This species was previously confused with *C. oblongiusculus. C. grimmi* has so far only been found in the immediate vicinity of the Red Sea.

VI.2.3.2.2 Clitobius oblongiusculus (Fairmaire, 1875) (see Plate XIX.1)

Halonomus oblongiusculus Fairmaire, 1875: 495.
Halonomus lineicollis Fairmaire, 1879a: 196, 1879b: 4.

Halonomus schneideri Allard, 1882: lxxxvi.
Material examined. Ouady Athal, 13.2.<19>02 (= Wadi eth-Thal) (MMHNP).
Referenced localities. Gebel Helal, El Arish (Alfieri 1976 sub *C. oblongiusculus* var. *lineicollis*), the Sinai (Alfieri 1976, f. t., Español 1967, Gridelli 1930 t. Peyerimhoff, Kaszab 1982, Peyerimhoff 1931, Reichardt 1936), Tor (Koch 1935a sub *C. oblongiusculus* ssp. *lineicollis*), Wadi eth-Thal (Peyerimhoff 1907 sub *C. oblongiusculus* var. *lineicollis*).
General distribution. Saharo-Turano-Sindian chorotype: Algeria up to Arabia and Tadzhikistan (Carl 1991b), Morocco (Kocher 1958), former Spanish Sahara, Mauritania (Español 1967), Sudan (Ardoin 1972a).
Comments. The synonymy of *C. oblongiusculus* and *C. oblongiusculus schneideri* was established by Purchart & Kamiński (2017). Specimens from Tor, Gebel Helal and El Arish were not seen by the authors. Halophile species (Peyerimhoff 1931).

VI.2.3.3   Genus *Proscheimus* Desbrochers des Loges, 1881

VI.2.3.3.1   Proscheimus arabicus Desbrochers des Loges, 1881

Proscheimus arabicus Desbrochers des Loges, 1881: 128.
Material examined. 20 km N Nuweiba, 3.XII.1992, L.P. (5); 2 km N Nuweiba, 28.XI.1992, L.P. (1) (ML).
Referenced locality. SW-Asiatic chorotype: The Sinai (Alfieri 1976 t. Peyerimhoff, Andres 1931b, Kaszab 1982, Reichardt 1936), Tor (Koch 1935a).
General distribution. Egypt, the Sinai, Jordan, Saudi Arabia (Kaszab 1982), Israel (Iwan et al. 2020).

VI.2.3.4   Genus *Diphyrrhynchus* Fairmaire, 1849

VI.2.3.4.1   Diphyrrhynchus aenescens (Fairmaire, 1892)

Abantis aenescens Fairmaire, 1892: 110.
Referenced localities. Isla (Alfieri 1976), Tor (Gridelli 1953, Koch 1935a, Schawaller 1991), the Sinai (Koch 1934).
General distribution. Red Sea seashore chorotype: West coast of Red Sea from Djibouti up to the Sinai, Yemeni coast, Camaran, Perim (Schawaller 1991), Saudi Arabia (Iwan et al. 2020).

VI.2.3.5   Genus *Gonocephalum* Solier, 1834

VI.2.3.5.1   Gonocephalum (Gonocephalum) patruele patruele (Erichson, 1843) (see Plate XIX.2)

Opatrum patruele Erichson, 1843: 248.
?*Hopatrum sericinum* Reiche (Walker 1871).
?*Hopatrum sericeum* Reiche (Innes Bey 1912).
Referenced localities. Wadi El Arish (Alfieri 1976), the Sinai (Andres 1931b).
General distribution. *G. patruele patruele*: Saharo-Turano-Sindian chorotype. Cape Verde Islands, Canary Islands, Libya, Egypt, Israel, Iran, Somalia, Sudan (Ferrer 2000b), Ethiopia (Ferrer 1995a), Niger, Chad (Pierre 1961b), the Sinai (Alfieri 1976), Morocco (Kocher 1958), Algeria (Théront & Hollande 1965) and Tunisia, United Arab Emirates, Afghanistan, Iraq, Oman, Saudi Arabia, Syria, Yemen (Iwan et al.

2020). – *G. patruele turchestanicum* Gridelli, 1948: 9: Afghanistan, Kazakhstan, Kirgisia, Tajikistan, Turkmenistan, Uzbekistan, Tibet (Iwan et al. 2020).
Comments. The presence of *G. patruele turchestanicum* in Egypt, as reported by Iwan et al. (2020), seems questionable.

VI.2.3.5.2   Gonocephalum (Gonocephalum) perplexum (Lucas, 1846)
 (see Plate XIX.3)
Opatrum perplexum Lucas, 1846: pl. 29.
Referenced localities. Wadi Isla (Alfieri 1976).
General distribution. Mediterraneo-Sindian chorotype: Western Sahara, Morocco, Algeria, Libya, Israel, Syria, Somalia, Saudi Arabia (Ferrer 2000b), Yemen (Kaszab 1982), Tunisia (Ferrer 1995a), Egypt, the Sinai (Alfieri 1976), Jordan (Katbeh-Bader 1996), Croatia, Greece, Italy (Lampedusa, Pantelleria), Spain, Cyprus, Israel (Iwan et al. 2020).

VI.2.3.5.3   Gonocephalum (Gonocephalum) rusticum (Olivier, 1811) (see
 Plate XIX.4)
Opatrum rusticum Olivier, 1811: 498.
Referenced localities. The Sinai (Crotch 1872 sub *Opatrum rusticum*).
General distribution. Centralasiatic-Mediterranean chorotype: Southern Europe, Canary Islands, the Azores, Madeira, Egypt, Israel (Ferrer 1995a), Jordan (Katbeh-Bader 1996), from Arabian Peninsula to central Asia and China (Iwan et al. 2020).

VI.2.3.5.4   Gonocephalum (Gonocephalum) setulosum setulosum
 (Faldermann, 1837) (see Plate XX.1)
Opatrum setulosum Faldermann, 1837: 60.
 Gonocephalum demaisonis Allard, 1883: 32.
Material examined. Wadi El Arish, 40 km SW Nakhl, 27.4.<19>68, coll. H. Schweiger (3) (TAU).
Referenced localities. Wadi Helal (Alfieri 1976), Wadi Isla (Alfieri 1976 sub *G. demaisoni*), Hadjar er-Reqab, Wadi eth-Thal, Wadi Gharandel, Ras abou Zenimeh, Wadi el Ain (Peyerimhoff 1907), the Sinai (Gridelli 1952, Kaszab 1982, Mifsud & Scupola 1998, Peyerimhoff 1931, Reichardt 1936).
General distribution. *G. setulosum setulosum*: Centralasiatic-Mediterranean-Sahelian chorotype: Turkmenistan, Kazakhstan, Caucasus, Iraq, Iran, Syria, Arabian Peninsula, Turkey ("Asia Minor"), Israel, Egypt, Sudan, Libya, Tunisia, Niger, Cyprus, Crete, Rhodos, Cyclades, Malta (Ferrer 1995a), Mauritania (Akjoujit) (ML, det. Ferrer). – *G. s. angustum* Lindberg, 1950: 11 (= *G. setulosum lindbergi* Ferrer, 1993: 75): Fuerteventura (Lindberg, 1950), Lower Egypt, Italy, Tunisia, Algeria, Morocco, western Sahara (Ferrer 1993, 1995a). – *G. setulosum kashgarense* (Bates, 1879): Iraq to China (Iwan et al. 2020).

VI.2.3.5.5   Gonocephalum (Gonocephalum) soricinum soricinum (Reiche &
 Saulcy, 1857) (see Plate XX.2)
Opatrum soricinum Reiche & Saulcy, 1857: 259.
 Material examined. Wadi Rim, 1.V.1999, P. (1) (TP).

Referenced localities. Wadi Isla, El Kontella (Alfieri 1976), Rinham, 1500 m (Andres 1931b), Wadi Sa'al, Ain el-Houdra (Andres 1931b t. Peyerimhoff), Wadi Isla (Bodenheimer & Theodor 1929), the Sinai (Español 1943, 1967, Gridelli 1939c, 1953, Peyerimhoff 1931, Reichardt 1936, Schuster 1938), Wadi Isla, Wadi Hebran (Koch 1935a), Wadi Sa'al, Ain el-Houdra (Peyerimhoff 1907).

General distribution. *G. soricinum soricinum*: Saharo-Sahelo-Sindian chorotype: Algeria, Benin, Chad, Erythrea, Ethiopia, Senegal, Sudan, India, Afghanistan, the Sinai, Iran, Syria, Israel, Jordan, Pakistan, Saudi Arabia, United Arab Emirates, Yemen (incl. Socotra) (Iwan et al. 2010). – *G. soricinum faraonicum* Ferrer, 2000: 74: Sudan, Ethiopia (Ferrer 2000b). – *G. soricinum insidiosum* (Fairmaire, 1880: 16): Morocco, Algeria (Ferrer 2000b).

VI.2.3.6 Genus *Mesomorphus* Miedel, 1880

VI.2.3.6.1 Mesomorphus setosus (Mulsant & Rey, 1853)

Opatrinus setosus Mulsant & Rey, 1853: 88.

Opatrum (Gonocephalum) murinum Baudi di Selve, 1876a: 227.

Referenced localities. El Kontella (Alfieri 1976), Oase Ain Musa (Andres 1920), Tor (Koch 1935a), garden of the Sinai monastery (Heyden 1899 sub *Opatrinus setosus*), S. Caterina (Gridelli 1939c t. Peyerimhoff sub *Mesomorphus murinus murinus*), Monastery of St. Catherine (Peyerimhoff 1907 t. Heyden).

General distribution. Levantine chorotype: Egypt, Israel (Ferrer 2000a), the Sinai (Alfieri 1976).

Comments. Zalat et al. (2008) report *Mesomorphus* spec. from the St Katherine Protectorate. It could be *M. setosus*.

VI.2.3.7 Genus *Opatroides* Brullé, 1832

VI.2.3.7.1 Opatroides punctulatus punctulatus Brullé, 1832

Opatroides punctulatus Brullé, 1832: 220

Material examined. 8 km S Taba, 26.XI.1992, L.P. (7); 10 km S Taba, 27.XI.1992, L.P. (8); 13 km S Taba, 3.XII.1992, L.P. (5); 14 km S Taba, 27.XI.1992, L.P. (8); 16 km S Taba, 27.XI.1992, L.P. (3); 19 km S Taba, 3.XII.1992, L.P. (3); 38 km S Taba, 3.XII.1992, L.P. (1); 22 km N Nuweiba, 3.XII.1992, L.P. (2); 3 km N Nuweiba, 3.XII.1992, L.P. (2); 3 km E Ain El Furtaga, 28.XI.1992, L.P. (1); 20 km S Nuweiba, 28.XI.1992, L.P. (16); 20 km NNW Dahab, 2.XII.1992, L.P. (18); Santa Katarina, 1400 m, 1.-2.XII.1992, L.P. (27); Mt. Moses (Gebel Musa), 7.-8.IV.1993, P. (5); 22 km NE Santa Katarina, 1300 m (in the garden), 2.XII.1992, L.P. (4); 27 km NE Santa Katarina, 1300 m, 2.XII.1992, L.P. (6); Dahab, 9.IV.1993, P. (1); 45 km SSW Dahab, 29.XI.1992, L.P. (1); 35 km N Sharm el Sheikh, 29.XI.1992, L.P. (53); 15 km N Sharm el Sheikh, 30.XI.1992, L.P. (8); Dahab, 21.4.1995, P. (1); Santa Katarina, 14.4.1995, P. (1); Santa Katarina, 16.4.1995, P. (3); Mt. Moses, 16.4.1995, P. (2) (ML, TP); area around oasis Ain Khudra, 25.-29.XII.1997, 600-800 m, leg. W. Schawaller (1); Sharm el Sheik around Shark's Bay, 1.-4.I.1998, leg. W. Schawaller (1) (SMNS).

Referenced localities. El Kontella, Wadi El Arish (Alfieri 1976), between Wadi Werdan and Wadi Feiran, between Tor and Djebel Hamam, desert of Ga'a, Wadi Gergir, Wadi Budr, Wadi Ta-albi, Wadi Chaschibi (Andres 1920), the Sinai (Crotch 1872 sub *Penthicus punctulatus*, Gebien 1906, Español 1943, Peyerimhoff 1931, Reichardt 1936), garden of the Sinai monastery (Heyden 1899 sub *Penthicus (Opatroides) punctulatus*),

Wadi Isla (Koch 1935a), Wadi Gharandel (Walker 1871, Peyerimhoff 1907 t. Walker), Gulf of Aqaba (Peyerimhoff 1907), St. Katherine Protectorate (Zalat et al. 2008).

General distribution. Mediterraneo-Sindian chorotype: Mediterranean region, the Caucasus, Central Asia, western Siberia, Eritrea, Somalia, Sudan, the Near East, Iraq, Iran, Afghanistan, Pakistan, India, and the Arabian Peninsula, introduced to the USA. – *O. punctulatus parvulus* (Faldermann, 1837) (= *subcylindricus* (Ménétriés, 1849)) is found in Azerbaijan, Armenia, Turkey, Iran, Oman, Tajikistan, Turkmenistan, Uzbekistan (Iwan et al. 2020, Kaszab 1982). – *O. punctulatus lilligi* Ferrer, 2005: 17: Mauritania, Niger, Sudan.

Comments. *O. punctulatus punctulatus* is significantly ($P = 0.02$) more prevalent at the South Sinai than at the North Sinai (Table II.1, Map VI.7).

VI.2.3.8   Genus *Scleropatroides* Löbl & Merkl, 2003

VI.2.3.8.1   Scleropatroides hirtulus (Baudi di Selve, 1875) (see Plate XX.3)
Opatrum hirtulum Baudi di Selve 1875: 703.

Referenced localities. Wadi Gharandel (Peyerimhoff 1907, Alfieri 1976 t. Peyerimhoff), the Sinai (Reichardt 1936), all sub *Scleropatrum hirtulum*.

General distribution. Centralasiatic-Mediterranean chorotype: Armenia, Transcaspian, Iran, Iraq, Syria, Afghanistan, Turkmenistan, Uzbekistan, Kazakhstan (Kaszab 1969), the Sinai (Reichardt 1936).

VI.2.3.9   Genus *Eurycaulus* Fairmaire, 1868

The subgenera mentioned by Koch (1935a) are questioned by Schawaller (1993). The subgenus *Scleronimon* Reitter, 1904 is a synonym of the genus *Ammotrypes* Fairmaire, 1879a which was first downgraded to subgenus of the genus *Eurycaulus* (Grimm 2001) and later regarded as a junior synonym of *Eurycaulus* (Grimm 2005).

VI.2.3.9.1   Eurycaulus (Eurycaulus) henoni Fairmaire, 1897 (see Plate XX.4)
Eurycaulus henoni Fairmaire, 1897: 121.
Eurycaulus (Scleronimon) peyerimhoffi Reitter, 1904: 127.

Referenced localities. Magdaba (Alfieri 1920), Magdaba, El Kontella (Alfieri 1976 sub *E. henoni* = *E. peyerimhoffi*), south of Ain Hawara (Andres 1920 sub *E. henoni* = *E. peyerimhoffi*), the Sinai (Andres 1921, 1931b sub *E. henoni* = *E. peyerimhoffi*, Reitter 1904, Schawaller 1993), El Arish (Koch 1935a), Hadjar er-Reqab (Peyerimhoff 1907).

General distribution. Southern Levantine chorotype: The Sinai (Schawaller 1993), the Negev (TAU).

Comments. For Peyerimhoff (1907), the type comparison leads to a synonymy of *E. peyerimhoffi* Reitter, 1904 with *E. henoni* Fairmaire, 1897. Koch (1935a) and Schawaller (1993) treat them as two different species. Grimm (2005) agrees with Peyerimhoff's opinion.

VI.2.3.10   Genus *Sclerum* Dejean, 1834

VI.2.3.10.1   Sclerum orientale orientale (Fabricius, 1775)
Opatrum orientale Fabricius, 1775: 76.

Referenced localities. Wadi Helal, Wadi Isla (Alfieri 1976), oasis of Feiran (Andres 1920), Wadi el Ain (Hart 1891, Peyerimhoff 1907 t. Hart), oasis of el-Hessoueh (Peyerimhoff 1907), the Sinai (Reichardt 1936).

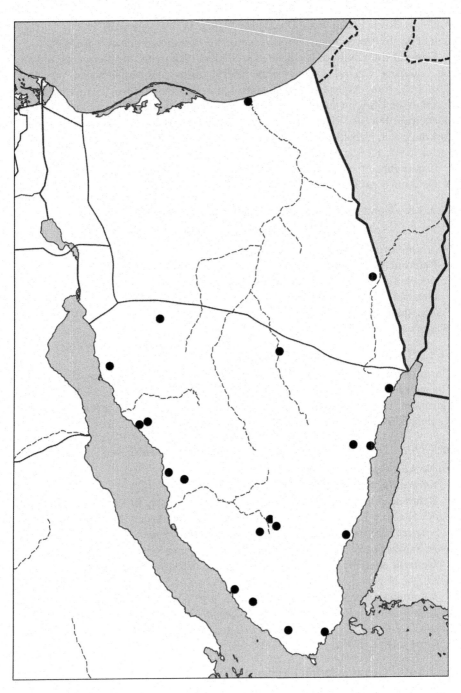

MAP VI.7 Distribution map of *Opatroides punctulatus punctulatus* in the Sinai Peninsula.

General distribution. Saharo-Sahelo-Arabian chorotype: Cameroon, Chad, Sudan, Ethiopia, Somalia, Saudi Arabia, Yemen (Kaszab 1982), Mauritania (Ardoin 1972b), Niger (Rougon & Ardoin 1976), Jordan (Ferrer 1995b), Palestine, Egypt (Pierre 1961b), the Sinai (Reichardt 1936). – *S. orientale yemense* Kaszab, 1972: Yemen.

VI.2.4   Tribe Pedinini Eschscholtz, 1829

VI.2.4.1   Genus *Drosochrus* Erichson, 1843

VI.2.4.1.1   Drosochrus (Helopinus) costatus aegyptiacus (Gridelli, 1939) (see Plate XXI.1)

Helopinus costatus aegyptiacus Gridelli, 1939a: 198.

Referenced localities. Wadi Isla (Alfieri 1976 sub *Helopinus costatus* ab. *elegans* Baudi di Selve),

Wadi Gneh, Wadi Taba (Gridelli 1939a, Peyerimhoff 1907 sub *Helopinus* sp.).

General distribution. *D. costatus costatus* Solier, 1848: Saudi Arabia, Jordan (Gridelli 1939a), Egypt, Afrotropis (Iwan et al. 2020). – *D. costatus aegyptiacus*: Lower Egypt, the Sinai (Gridelli 1939a), Afrotropis (Iwan et al. 2020), Israel (ML). – *D. costatus elegans* (Baudi di Selve, 1881: 287): Eritrea, Djibouti (Gridelli 1939a).

Comments. The occurrence of *D. costatus aegyptiacus* in the Afrotropis seems questionable.

VI.2.4.2   Genus *Leichenum* Dejean, 1834

VI.2.4.2.1   Leichenum pulchellum pumilum Baudi di Selve, 1876 (see Plate XXI.2)

Leichenum pulchellum (?) var. *pumilum* Baudi di Selve, 1876b: 73.

Referenced localities. Wadi Feiran, Wadi El Ain (Alfieri 1976, Peyerimhoff 1907 sub *Leichenum pulchellum* Lucas, Andres 1931b sub *Leichenum pulchellum* Küster), Wadi Feiran (Koch 1935a sub *L. pulchellum* Küster), the Sinai (Español 1973, Gridelli 1939b, Peyerimhoff 1931 sub *L. pulchellum* Küster).

General distribution. *L. pulchellum pulchellum* (Lucas, 1846: 336): western European and African Mediterranean. – *L. pulchellum pumilum*: Mediterraneo-Sindian chorotype: Eritrea, Egypt, the Sinai, Lebanon, Turkey ("Asia Minor"), Cyprus, Iraq, the Arabian Peninsula, Bahrain (Gridelli 1939b, Kaszab 1982), Chad (Español 1973), Azerbaijan, Greece, Libya, Tunisia, Iran, Israel, Pakistan, Syria (Iwan et al. 2020).

VI.3   Subfamily Tenebrioninae Latreille, 1802

VI.3.1   Tribe Alphitobiini Reitter, 1917

VI.3.1.1   Genus *Alphitobius* Stephens, 1829

VI.3.1.1.1   Alphitobius diaperinus (Panzer, 1796) (see Plate XXI.3)

Tenebrio diaperinus Panzer, 1796: 16.

Material examined. Nuweiba, IV.1992, Ullrich (1) (CGW).

Referenced localities. The Sinai (Andres 1931b), probably Wadi Tarfa (Kneucker 1903).

General distribution. Subcosmopolitan (Mifsud & Scupola 1998).

VI.3.2 Tribe Helopini Latreille, 1802

VI.3.2.1 Genus *Hedyphanes* Fischer von Waldheim, 1820

VI.3.2.1.1 Hedyphanes (Microhedyphanes) chikatunovi Nabozhenko & Lillig, 2013 (see Plate XXI.4)

Hedyphanes (Microhedyphanes) chikatunovi Nabozhenko & Lillig, 2013: 188.

Referenced localities. 90 km east of Suez Canal, airfield Bir Gifgafa, 30°24'26" N, 33°9'15" E (Nabozhenko & Lillig 2013).

General distribution. Southern Levantine chorotype: Israel (Negev), the Sinai.

Comments. So far to the authors only known from Israel and the Sinai Peninsula and not from the Lower Egypt, as mentioned in Iwan et al. (2020).

VI.3.3 Tribe Melanimonini Seidlitz, 1894

VI.3.3.1 Genus *Cheirodes* Gené, 1839

VI.3.3.1.1 Cheirodes (Cheirodes) sardous sardous Gené, 1839

Cheirodes sardous Gené, 1839: 73.

Referenced localities. Wadi Isla up to vicinity of Monastery of St. Catherine (Alfieri 1976 sub *Anemia sardoa*), between Wadi Werdan and Wadi Feiran (Andres 1920 sub *Anemia sardoa*), the Sinai (Andres 1931b, Gridelli 1933a sub *Anemia sardoa*, Reichardt 1936 sub *Anemia sardoa*), Wadi Gneh, Wadi Feiran (Peyerimhoff 1907 sub *Anemia sardoa*).

General distribution. *Ch. sardous sardous*: Canary Islands, southern Europe up to Iran (Ardoin 1971a), introduced in Australia (Matthews 2004). *Cheirodes sardous denticulatus* (Wollaston, 1867: 197): Cape Verde Islands up to Ethiopia (Ardoin 1971a), Yemen (Kaszab 1982).

Comments. Ferrer (2016) revised the *sardous* group. In doing so, he split *Ch. sardous* into several species. This revision is controversial. Three of the nine new species have already been retracted as synonyms to *Ch. sardous* (Castro Tovar 2020). Iwan et al. (2020) ignore Ferrer's paper. Among the new species is *Ch. linnmani* Ferrer, 2016 from Israel. With Eilat a locality close to the Sinai is mentioned. Also mentioned is "Gaza", but the locality is wrong. The animals in question (paratypes, ML) bear the labels "Israel: Golan Heights, Hat Ness, 8-15.VII.1996, leg. G. Sana" (2 ex.) and "Israel, Carmel, Basmat Ta'bun, 32°44'N 35°09'E, 4.V.2004, oak bark, leg. T. Pavlíček" (1 ex.).

Ferrer (2016) attributes the records cited by Lillig & Pavlíček (2003) to his *Ch. linnmani* Ferrer, 2016 without having seen these specimens. Because of these uncertainties we follow Iwan et al. (2020) here and stick for the time being with *Cheirodes sardous sardous* as a name of the specimens from Sinai.

VI.3.3.1.2 Cheirodes (Histiaea) asperulus (Reitter, 1884) (see Plate XXII.1)

Anemia asperula Reitter, 1884: 260.

Anemia asperula var. *seriesetosa* Baudi di Selve, 1894: 8.

Anemia denticulata Pic, 1923a: 5 nec Wollaston, 1867: 197.

Anemia aegyptiaca Pic, 1936: 300, replacement name.

Referenced localities. Wadi Gederat, Wadi Helal, Wadi Isla (Alfieri 1976 sub *Anemia aegyptiaca*), between Wadi Mezara and Um Shousha (Alfieri 1920 sub

Anemia (?) *seriesetosa*), Wadi Mezara, Um Shousha (Andres 1931b sub *Anemia denticulata*).
General distribution. Saharo-Sahelo-Sindian chorotype: South of Sahara from Senegal to Sudan, Egypt, Israel, Syria, Jordan, Iran (Ardoin 1971a), Saudi Arabia, Yemen (Iwan et al. 2020).

VI.3.3.1.3 Cheirodes (Pseudanemia) brevicollis (Wollaston, 1864) (see Plate XXII.2)
Pseudanemia brevicollis Wollaston 1864: 493.
  Anemia aphodioides Walker, 1871: 16.
  Anemia fausti Solsky, 1881: 52.
  Anemia pharao Reitter, 1897: 43.
Referenced localities. Wadi Mezara, Um Shusha (Alfieri 1920, 1976 sub *Anemia fausti*), Wadi Budhr, Mear (Andres 1931b sub *Anemia pharao*), Wadi eth Thal (Andres 1931b t. Peyerimhoff, Peyerimhoff 1907), the Sinai (Ardoin 1971a, Gridelli 1933a sub *Anemia fausti*, Gridelli 1952 sub *Anemia brevicollis*, Peyerimhoff 1931 sub *Anemia brevicollis*, Reichardt 1936 sub *Anemia aphodioides*), Wadi Fawar (Koch 1935a sub *Anemia fausti*), Wadi Gneh, Wadi Feiran (Peyerimhoff t. Walker sub *Anemia* (?) *aphodioides*), Wadi Genneh, Wadi Feiran (Walker 1871 sub *Anemia aphodioides* sp. n.).
General distribution. Centralasiatic-Mediterranean chorotype: Canary Islands, North Africa, eastern Africa, southern Europe, Arabian Peninsula, Near East, central Asia (Kaszab 1982), China (Iwan et al. 2020).

VI.3.3.1.4 Cheirodes (Spinanemia) pilosus (Tournier, 1868) (see Plate XXII.3)
Anemia pilosa Tournier, 1868: 146.
Referenced localities. Vicinity of Monastery of St. Catherine (Alfieri 1976 sub *Anemia pilosa*), desert of Ga'a (Andres 1920 sub *Anemia pilosa*), the Sinai (Gridelli 1952 t. Andres sub *Anemia pilosa*).
General distribution. Saharo-Sahelo-Arabian chorotype: Morocco up to Red Sea (Ardoin 1971a), Israel (Negev) (Ayal & Merkl 1994), Saudi Arabia, Oman, Yemen, the Sinai, Mauritania, Niger, Chad, Sudan (Kaszab 1982), Canary Islands (Oromí 1982).

VI.3.4 Tribe Scaurini Billberg, 1820

VI.3.4.1 Genus *Scaurus* Fabricius, 1775

VI.3.4.1.1 Scaurus aegyptiacus Solier, 1838
Scaurus aegyptiacus Solier, 1838b: 170.
  Scaurus giganteus Küster, 1848: 54.
Referenced localities. Wadi Isla, Wadi Helal (Alfieri 1976), between Wadi Werdan and Wadi Feiran (Andres 1920), the Sinai (Gridelli 1930 t. Andres, Peyerimhoff 1948b), Nachle (Heyden 1899), Qala'at en-Nakhel (Peyerimhoff 1907 t. Heyden).
General distribution. Mediterraneo-Sindian chorotype: Malta, Greece, Italian islands, Cyprus, Morocco to Egypt, Iraq, Lebanon, Syria, Turkey (Iwan et al. 2020) and Sudan (Lillig 1995).

VI.3.4.1.2 Scaurus rugicollis Reitter, 1914

Scaurus puncticollis (?) v. *rugicollis* Reitter, 1914c: 374.

Referenced localities. Wadi Isla (Alfieri 1976), Ain Musa, between Wadi Feiran and Wadi Selaf (Andres 1920), El Arish, Tor (Andres 1931b), the Sinai (Gridelli 1939d t. Andres 1920 and Andres, 1931b, Koch 1965, Peyerimhoff 1948b).

General distribution. Levantine chorotype: Israel/Palestine, the Sinai (Ferrer et al. 2014).

Comments. The *puncticollis* group was revised by Ferrer et al. (2014) and the species of the group occurring in Turkey by Nabozhenko et al. (2020). According to Ferrer et al. (2014), all records from the Sinai refer to *Sc. rugicollis* and not to *Sc. puncticollis*. *Rugicollis* is listed as a synonym of *Sc. puncticollis* Solier, 1838b: 172 by Iwan et al. (2020), who ignore this.

VI.3.5   Tribe Triboliini Gistel, 1848

VI.3.5.1   Genus *Tribolium* MacLeay, 1825

VI.3.5.1.1   Tribolium (Tribolium) castaneum (Herbst, 1797) (see Plate XXII.4)

Colydium castaneum Herbst, 1797: 282.

Dermestes navalis Fabricius, 1775: 56 (suppressed name).

Referenced localities. Wadi Isla, Wadi Feiran, El Arish, El Kantella, Wa Helal, Megara (Alfieri 1976), Wadi Gergir (Andres 1920), Feiran, in dates (Peyerimhoff 1907 sub *Tribolium navale* Fabricius).

General distribution. Cosmopolitan, frequently synanthropic.

Comments. The name "*Dermestes navalis* Fabricius, 1775" has been suppressed, and the name *castaneum* (Herbst, 1797) conserved (Pope & Watt 1986).

VI.3.5.1.2   Tribolium (Stene) confusum Jacquelin du Val, 1861
            (see Plate XXIII.1)

Tribolium confusum Jacquelin du Val, 1861: 181.

Referenced localities. Wadi Isla, Monastery of Sainte-Catherine (Alfieri 1976).

General distribution. Cosmopolitan, frequently synanthropic.

VI.4   Subfamily Diaperinae Latreille, 1802

VI.4.1   Tribe Crypticini, Brullé, 1832

VI.4.1.1   Genus *Pseudoseriscius* Español, 1950

VI.4.1.1.1   Pseudoseriscius maculosus murinus (Allard, 1882)

Crypticus murinus Allard, 1882: lxxxvii

Referenced localities. El Qantara (Boyd 1917 sub *Crypticus murinus*).

General distribution. *P. maculosus maculosus* Fairmaire, 1870: "Syria", Lebanon, Israel (Español 1949), Iran (Bushir) (Gebien & Borchmann 1927), Saudi Arabia (Iwan et al. 2020). – *P. maculosus murinus*: Southern Levantine chorotype: Lower Egypt (Español 1949), the Sinai (Boyd 1917).

Comments. Reports from Libya and former Spanish Sahara must be incorrect (Español 1949).

VI.4.2 Tribe Phaleriini Blanchard, 1845

VI.4.2.1 Genus *Phaleria* Latreille, 1802

VI.4.2.1.1 Phaleria prolixa Fairmaire, 1869 (see Plate XXIII.2)

Phaleria prolixa Fairmaire, 1869: 797.
  Phaleria munda Walker, 1871: 16.
  Phaleria aegyptiaca Seidlitz, 1894: 482.
  Material examined. 10 km S Taba, 27.XI.1992, L.P. (1 var. *aegyptiaca*); 3 km N Nuweiba, 3.XII.1992, L.P. (5 forma *typica*, 10 var. *aegyptiaca*); Sharm el Sheikh, 30.XI.1992, L.P. (4 var. *aegyptiaca*); Dahab, 20.4.1995, P. (1); Dahab, 21.4.1995, P. (3); Dahab, 22.4.1995, P. (3) (all var. *aegyptiaca*) (ML, TP).
  Referenced localities. Rafa, El Arish (Alfieri 1976), the Sinai (Andres 1931b, Gebien & Borchmann 1927, Reitter 1916b), Tor (Gridelli 1953 t. Walker 1871, Innes Bey 1912), Ras abou Zenimeh (Gridelli 1953 t. Peyerimhoff, Peyerimhoff 1907 sub *P. munda*), Tor (Peyerimhoff 1907 t. Walker, Walker 1871 sub *Ph. munda* sp. n.).
  General distribution. NE-African-Sindian chorotype: Coasts of the Red Sea and of Arabia (Schawaller 1993), Iran (Bushir) (Gebien & Borchmann 1927), Somalia, western coast of Madagascar, probably also eastern Africa (Ardoin 1974), Mediterranean coast of Egypt (Alfieri 1976).
  Comments. According to Kaszab (1981), the species lives on algae in the sand of the coast. The authors found it near Nuweiba and Sharm el Sheikh in washed up dead fish and near Taba on a dead crab. The nominal form and the variety *aegyptiaca* live together. All kinds of transitions exist between them.

VI.4.2.2 Genus *Phtora* Germar, 1836

VI.4.2.2.1 Phtora (Phtora) apicilaevis (Marseul, 1876)

Cataphronetis apicilaevis Marseul, 1876: 36.
  Material examined. 5 km N Nuweiba, 28.XI.1992, L.P. (1) (ML).
  Referenced localities. El Arish, Wadi Karam (Alfieri 1976 sub *Cataphronetis apicilaevis*).
  General distribution. Libya, Lower Egypt (Gridelli 1930), the Sinai (Alfieri 1976), Tunisia (Normand 1936), Sudan (Bunalski & Śmiełowski 2018).
  Comments. See *Phtora subclavata*.

VI.4.2.2.2 Phtora (Phtora) subclavata (Wollaston, 1861) (see Plate XXIII.3)

Pseudostene subclavata Wollaston, 1861: 250.
  Referenced localities. El Arish (Alfieri 1976 sub *Pseudostena subclavata*).
  General distribution. Egypt, Saudi Arabia (Kaszab 1982), the Sinai (Alfieri 1976), Israel (Iwan et al. 2020).
  Comments. According to Grimm (1995) *P. subclavata* (Wollaston, 1861), *P. apicilaevis* (Marseul, 1876) and other "species" are probably identical to *Phtora angusta* (Wollaston, 1861).

VI.4.3 Tribe Trachyscelini Blanchard, 1845

VI.4.3.1 Genus *Trachyscelis* Latreille, 1809

VI.4.3.1.1 Trachyscelis tenuestriatus Fairmaire, 1886 (see Plate XXIII.4)
Trachyscelis tenuestriatus Fairmaire, 1886: 449.
Material examined. 3 km W Dahab, 3.XI.1992, L.P. (1) (ML).
Referenced localities. The Sinai (Andres 1931b t. Peyerimhoff, Kaszab 1982), Tor (Gridelli 1953 t. Koch 1935a), Gulf of Aqaba (Gridelli 1953 t. Peyerimhoff 1907, Koch 1935a), Hayeh (Gulf of Aqaba) (Peyerimhoff 1907).
General distribution. Red Sea seashore chorotype: Djibouti, Perim Island, Yemen, the Sinai, Jordan (Kaszab 1982), Sudan (CB).
Comments. Psammobiont species of the coast.

VI.5 Subfamily Alleculinae Laporte de Castelnau, 1840

VI.5.1 Tribe Alleculini Laporte de Castelnau, 1840

VI.5.1.1 Genus *Mycetocharina* Seidlitz, 1890

VI.5.1.1.1 Mycetocharina (Mycetocharina) megalops megalops (Fairmaire, 1894)
Caristela megalops Fairmaire, 1894: 311.
Mycetocharina longipennis Pic, 1923b: 10.
Referenced localities. Sinai (Pic, 1923b sub *M. longipennis*), St Katherine Protectorate (Zalat et al. 2008).
General distribution. *M. megalops megalops*: Saharian chorotype: Algeria, Egypt, Morocco, Tunisia, the Sinai, Israel, – *M. megalops pallidior* Pic, 1938: Morocco (Iwan et al. 2020).

VI.5.2 Tribe Cteniopodini, Seidlitz, 1896

VI.5.2.1 Genus *Cteniopus* Solier, 1835

VI.5.2.1.1 Cteniopus (Cteniopsomus) priesneri luteiceps Pic, 1939
Cteniopus priesneri var. *luteiceps* Pic, 1939: 146.
Referenced localities. Wadi el Arish (Pic 1939).
General distribution. *C. priesneri priesneri* Wittmer, 1936: Egypt, – *C. priesneri luteiceps*. Southern Levantine chorotype: Egypt, Sinai (Iwan et al. 2020).

VI.5.2.2 Genus *Omophlus* Dejean, 1834

VI.5.2.2.1 Omophlus (Omophlus) gracilipes Kirsch, 1869
Omophlus gracilipes Kirsch, 1869: 108.
Referenced localities. Wadi el Arish (Heyden 1899, Peyerimhoff 1907).
General distribution. SW-Asiatic chorotype: Egypt, Israel, Jordan, Syria, Turkey (Iwan et al. 2020), the Sinai (Heyden 1899).

Plate I
Photos 1–4

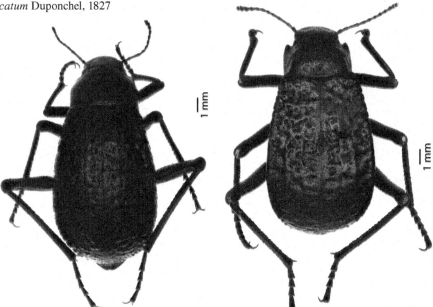

1 – *Adelostoma (Adelostoma) sulcatum sulcatum* Duponchel, 1827

2 – *Machlopsis (Machlopsis) crenatocostata* (Redtenbacher, 1868)

3 – *Adesmia (Adesmia) montana montana* (Klug, 1830)

4 – *Adesmia (Adesmia) cancellata abbreviata* (Klug, 1830)

All photographs by Edgar Müller, Germany, © Edgar Müller

Plate II
Photos 1–4

1 – *Adesmia (Oteroscelis) cothurnata cothurnata* (Forskål, 1775)

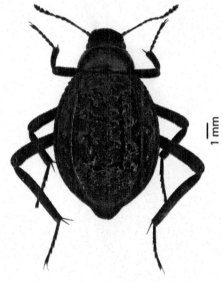

2 – *Adesmia (Oteroscelis) metallica brozai* Ardoin, 1978

3 – *Adesmia (Oteroscelis) metallica laevior* Ardoin, 1978

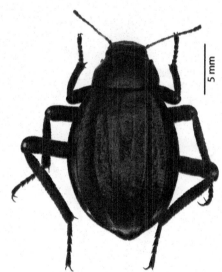

4 – *Adesmia (Oteroscelopsis) dilatata dilatata* (Klug, 1830)

All photographs by Edgar Müller, Germany, © Edgar Müller

Plate III
Photos 1–4

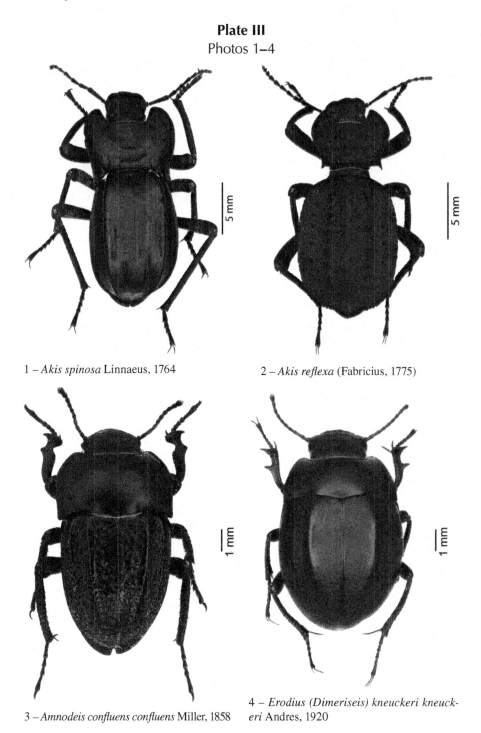

1 – *Akis spinosa* Linnaeus, 1764

2 – *Akis reflexa* (Fabricius, 1775)

3 – *Amnodeis confluens confluens* Miller, 1858

4 – *Erodius (Dimeriseis) kneuckeri kneuckeri* Andres, 1920

All photographs by Edgar Müller, Germany, © Edgar Müller

Plate IV
Photos 1–4

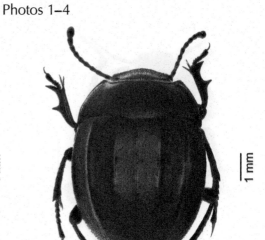

1 – *Erodius (Dirosis) dejeanii* Solier, 1834

2 – *Erodius (Dirosis) hebraicus* Lillig, 1997

3 – *Erodius (Dirosis) puncticollis sinaiticus* Crotch, 1872

4 – *Erodius (Erodius) gibbus gibbus* Fabricius, 1775

All photographs by Edgar Müller, Germany, © Edgar Müller

Plate V
Photos 1–4

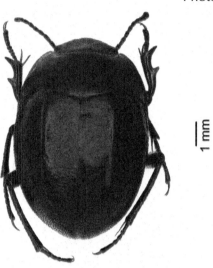

1 – *Erodius (Zophoserodius) zophosoides zophosoides* Allard, 1864

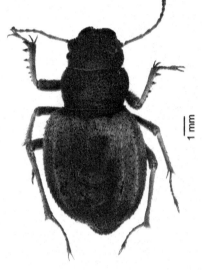

2 – *Astorthocnemis becvarorum* Lillig & Pavlíček, 2002

3 – *Pimelia (Pimelia) angulata angulata* Fabricius, 1775

4 – *Pimelia (Pimelia) arabica edomita* Koch, 1940

All photographs by Edgar Müller, Germany, © Edgar Müller

Plate VI
Photos 1–4

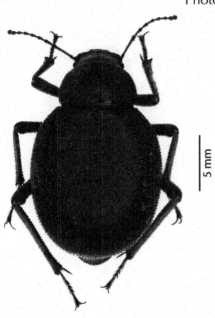

1 – *Pimelia (Pimelia) barthelemyi* Solier, 1836

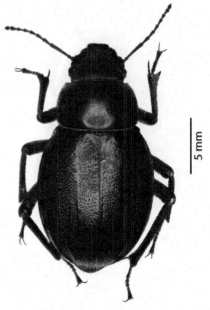

2 – *Pimelia (Pimelia) canescens canescens* Klug, 1830

3 – *Pimelia (Pimelia) mittrei* Solier, 1836

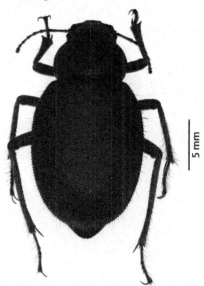

4 – *Pimelia (Pimelia) subquadrata subquadrata* Sturm, 1826

All photographs by Edgar Müller, Germany, © Edgar Müller

Plate VII
Photos 1–4

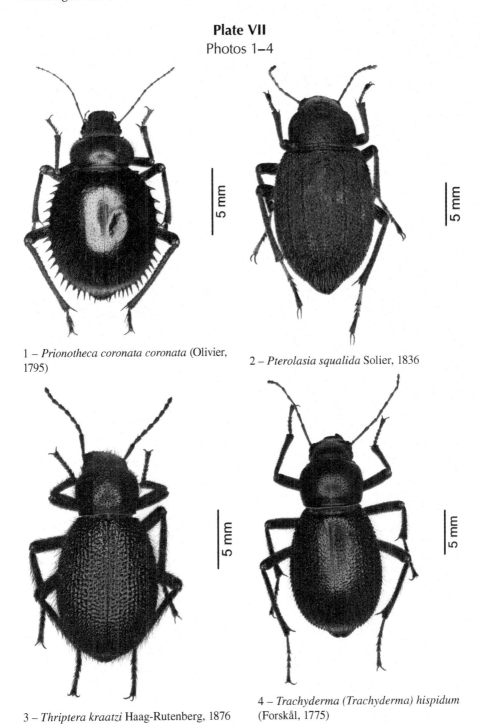

1 – *Prionotheca coronata coronata* (Olivier, 1795)

2 – *Pterolasia squalida* Solier, 1836

3 – *Thriptera kraatzi* Haag-Rutenberg, 1876

4 – *Trachyderma (Trachyderma) hispidum* (Forskål, 1775)

All photographs by Edgar Müller, Germany, © Edgar Müller

Plate VIII
Photos 1–4

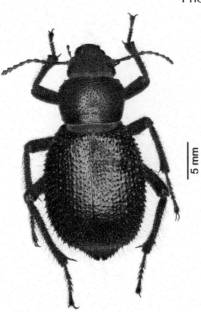

1 – *Trachyderma (Trachyderma) philistinum* Reiche & Saulcy, 1857

2 – *Sepidium dathan* Crotch, 1872

3 – *Sepidium tricuspidatum tricuspidatum* Fabricius, 1775

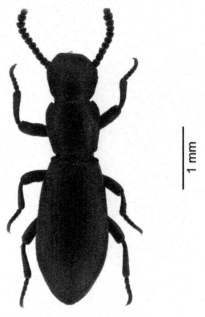

4 – *Dichillus (Dichillus) alfierii* Koch, 1935

All photographs by Edgar Müller, Germany, © Edgar Müller

Plate IX
Photos 1–4

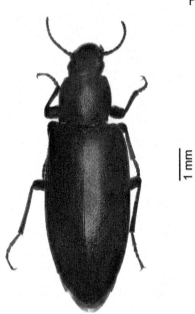

1 – *Cyphostethe (Cyphostethoides) heydeni* (Haag-Rutenberg, 1877)

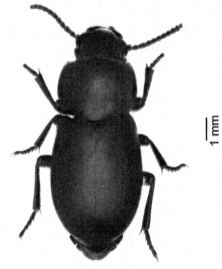

2 – *Dailognatha crenata* Reiche & Saulcy, 1857

3 – *Hegeterocara arabica* Reitter, 1900

4 – *Hionthis tentyrioides* Miller, 1861

All photographs by Edgar Müller, Germany, © Edgar Müller

Plate X
Photos 1–4

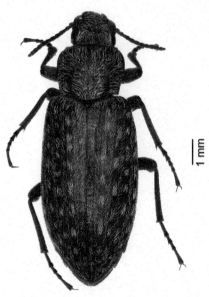

1 – *Imatismus villosus* (Haag-Rutenberg, 1870)

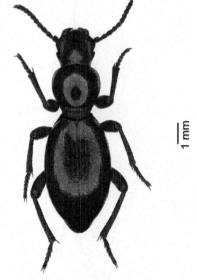

2 – *Mesostena (Mesostena) angustata angustata* (Fabricius, 1775)

3 – *Mesostena (Mesostena) puncticollis* Solier, 1835

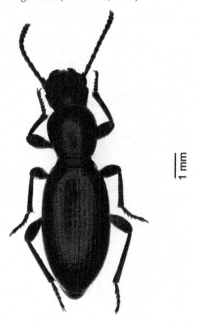

4 – *Mesostena (Mesostenopa) picea sinaitica* (Schatzmayr & Koch, 1934)

All photographs by Edgar Müller, Germany, © Edgar Müller

Plate XI
Photos 1–4

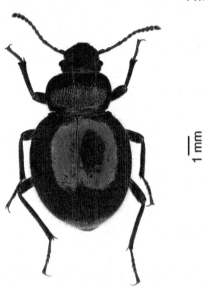

1 – *Micipsa (Micipsa) burtoni* Baudi di Selve, 1881

2 – *Micipsa (Micipsa) douei douei* Lucas, 1856

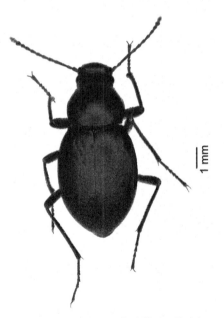

3 – *Micipsa (Micipsa) cf. philistina* Reiche & Saulcy, 1857

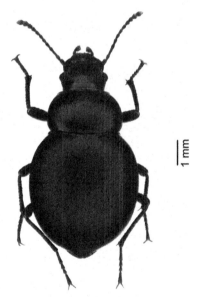

4 – *Micipsa (Micipsa) schaumii* Kraatz, 1865

All photographs by Edgar Müller, Germany, © Edgar Müller

Plate XII
Photos 1–4

1 – *Oterophloeus alveatus peyerimhoffi* Koch, 1935

2 – *Oterophloeus haagii orientalis* Koch, 1935

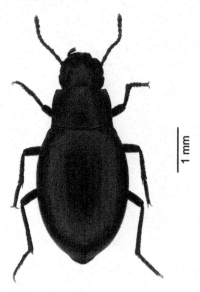

3 – *Oxycara (Oxycara) ardoini* Kaszab, 1979

4 – *Oxycara (Oxycara) pygmaeum* (Reiche & Saulcy, 1857)

All photographs by Edgar Müller, Germany, © Edgar Müller

Plate XIII
Photos 1–4

1 – *Oxycara (Symphoxycara) peyerimhoffi* Lillig, 2001

2 – *Prochoma (Prochoma) audouini* Solier, 1835

3 – *Scelosodis castaneus castaneus* (Eschscholtz, 1831)

4 – *Tentyria discicollis* Reiche & Saulcy, 1857

All photographs by Edgar Müller, Germany, © Edgar Müller

Plate XIV
Photos 1–4

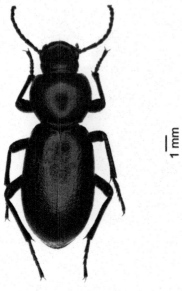

1 – *Tentyria sinaitica* Peyerimhoff, 1907

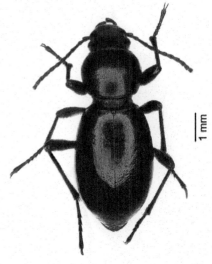

2 – *Tentyrina orbiculata subsulcata* (Reiche & Saulcy, 1857)

3 – *Tentyrina palmeri palmeri* (Crotch, 1872)

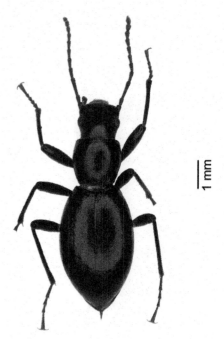

4 – *Thraustocolus (Thraustocolus) leptoderus* (Kraatz, 1865)

All photographs by Edgar Müller, Germany, © Edgar Müller

Plate XV
Photos 1–4

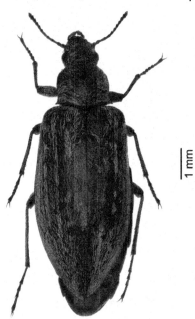

1 – *Trichosphaena perraudierei perraudierei* (Marseul, 1867)

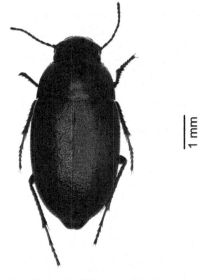

2 – *Zophosis (Hologenosis) nigroaenea* Deyrolle, 1867

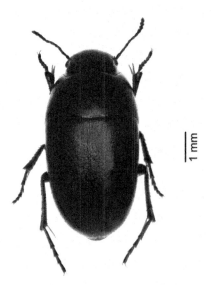

3 – *Zophosis (Oculosis) punctata punctata* Brullé, 1832

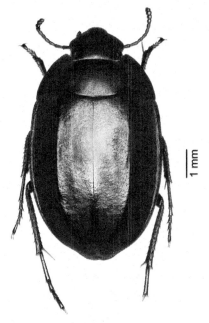

4 – *Zophosis (Septentriophosis) bicarinata* cf. *ghilianii* Deyrolle, 1867

All photographs by Edgar Müller, Germany, © Edgar Müller

Plate XVI
Photos 1–4

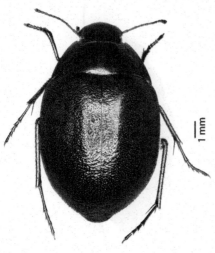

1 – *Zophosis (Septentriophosis) complanata* Solier, 1834

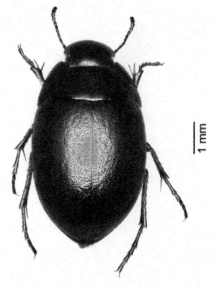

2 – *Zophosis (Septentriophosis)* lethierryi Deyrolle, 1867

3 – *Zophosis (Septentriophosis) personata* Erichson, 1841

4 – *Zophosis (Septentriophosis) pharaonis pharaonis* Reitter, 1916

All photographs by Edgar Müller, Germany, © Edgar Müller

Plate XVII
Photos 1–4

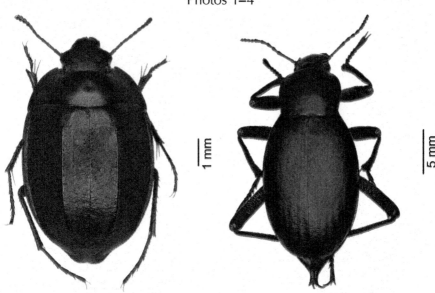

1 – *Zophosis (Septentriophosis) plana* (Fabricius, 1775)

2 – *Blaps bifurcata bifurcata* Solier, 1848

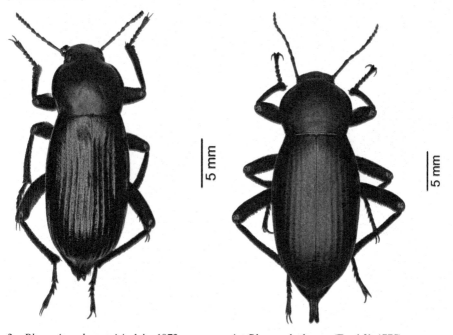

3 – *Blaps nitens laportei* Ardoin, 1973

4 – *Blaps polychresta* (Forskål, 1775)

All photographs by Edgar Müller, Germany, © Edgar Müller

Plate XVIII
Photos 1–4

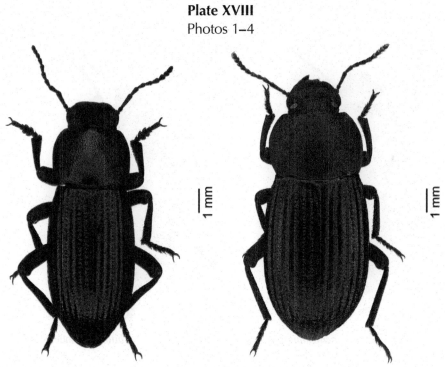

1 – *Dendarus (Rhizalemus) calcaroides* (Peyerimhoff, 1907)

2 – *Dendarus (Rhizalus) syriacus* (Reiche & Saulcy, 1857)

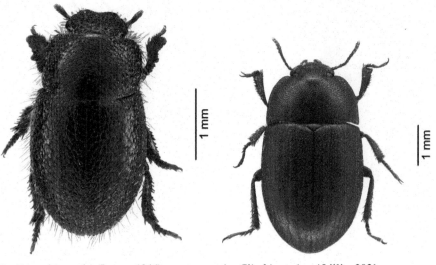

3 – *Ammobius rufus* (Lucas, 1846)

4 – *Clitobius grimmi* Lillig, 2021

All photographs by Edgar Müller, Germany, © Edgar Müller

Plate XIX
Photos 1–4

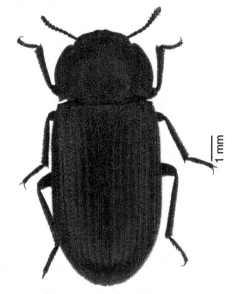

1 – *Clitobius oblongiusculus* (Fairmaire, 1875)

2 – *Gonocephalum (Gonocephalum) patruele patruele* (Erichson, 1843)

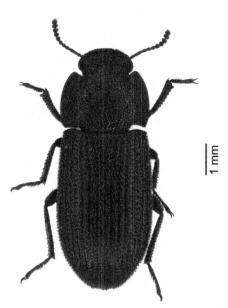
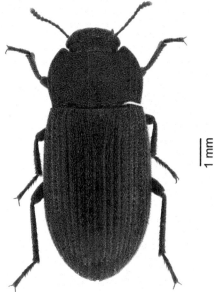

3 – *Gonocephalum (Gonocephalum) perplexum* (Lucas, 1846)

4 – *Gonocephalum (Gonocephalum) rusticum* (Olivier, 1811)

All photographs by Edgar Müller, Germany, © Edgar Müller

Plate XX
Photos 1–4

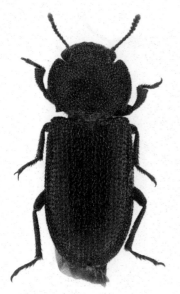

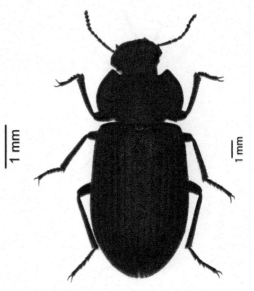

1 – *Gonocephalum (Gonocephalum) setulosum setulosum* (Faldermann, 1837)

2 – *Gonocephalum (Gonocephalum) soricinum soricinum* (Reiche & Saulcy, 1857)

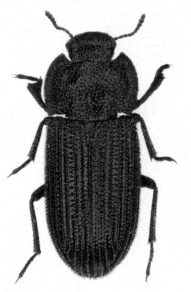

3 – *Scleropatroides hirtulus* (Baudi di Selve, 1875)

4 – *Eurycaulus (Eurycaulus) henoni* Fairmaire, 1897

All photographs by Edgar Müller, Germany, © Edgar Müller

Plate XXI
Photos 1–4

1 – *Drosochrus (Helopinus) costatus aegyptiacus* (Gridelli, 1939)

2 – *Leichenum pulchellum pumilum* Baudi di Selve, 1876

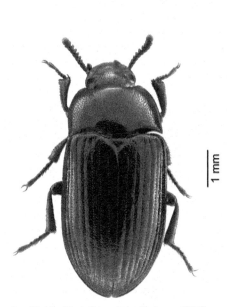

3 – *Alphitobius diaperinus* (Panzer, 1796)

4 – *Hedyphanes (Microhedyphanes) chikatunovi* Nabozhenko & Lillig, 2013

Photograph Nos. 1, 2, and 3 by Edgar Müller, Germany, © Edgar Müller.
Photograph No. 4 by M. Lillig.

Plate XXII
Photos 1–4

1 – *Cheirodes (Histiaea) asperulus* (Reitter, 1884)

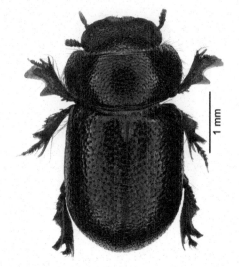

2 – *Cheirodes (Pseudanemia) brevicollis* (Wollaston, 1864)

3 – *Cheirodes (Spinanemia) pilosus* (Tournier, 1868)

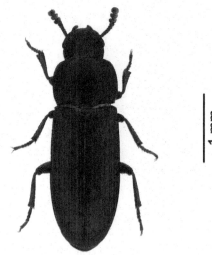

4 – *(Tribolium) castaneum* (Herbst, 1797)

All photographs by Edgar Müller, Germany, © Edgar Müller

Plate XXIII
Photos 1–4

1 – *Tribolium (Stene) confusum* Jacquelin du Val, 1868

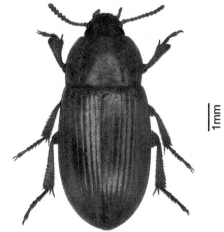

2 – *Phaleria prolixa* Fairmaire, 1869

3 – *Phtora (Phtora) subclavata* (Wollaston, 1861)

4 – *Trachyscelis tenuestriatus* Fairmaire, 1886

All photographs by Edgar Müller, Germany, © Edgar Müller

VII Incorrect and Doubtful Records (In Alphabetical Order)

1. *Adelostoma (Zarudnionymus) grande* **Haag-Rutenberg, 1879**

Adelostoma grande Haag-Rutenberg, 1879: 294.
Referenced localities. Wadi Feiran, Djebel Musa (Andres 1931b sub *Adelostoma grande* t. Peyerimhoff, Peyerimhoff 1907 sub *Adelostoma grande*), the Sinai (Koch 1935a sub *Adelostoma grande*).
General distribution. Iraq (Carl 1991a), Palestine, the Sinai (Koch 1935a).
Comments. Carl (1991) described the genus *Adelostomoides* with *Adelostoma grande* Haag-Rutenberg as the type species. Purchart (2007: 240) established its synonymy with *Adelostoma* subgenus *Zarudnionymus* Semenov-Tjan-Shansky & Bogatchev, 1947.

2. *Adesmia (Macradesmia) cancellata cancellata* **(Klug, 1830)**

Pimelia cancellata Klug, 1830: No 37.
Comments. Already Andres (1931b) questions its presence in the Sinai. It is certainly *A. cancellata latreillei* Solier.

3. *Adesmia (Oteroscelis) carinata carinata* **Solier, 1835**

Adesmia carinata Solier, 1835b: 547.
Referenced localities. The Sinai (Hart 1891, Peyerimhoff 1907 t. Hart).
General distribution. Syria, Iraq, Iran (Carl 1990), Jordan, Saudi Arabia, Turkey (Iwan et al. 2020).
Comments. The species cannot occur in the Sinai (Andres 1931b).

4. *Adesmia gracilis* **Gory**

Referenced localities. Mount Sinai (Walker 1871).
Comments. There is no *Adesmia* bearing this name (see Peyerimhoff 1907).

5. *Adesmia (Oteroscelis) lacunosa* **(Klug, 1830)**

Pimelia lacunosa Klug, 1830: No 33.
Referenced localities. Djebel Watiyeh (Hart 1891), Pharaoh's Baths (= Hammam Fir'aoun), Wadi Gharandel (Peyerimhoff 1907 t. Walker, Walker 1871), Djebel Ouatiyet (Peyerimhoff 1907 t. Hart), the Sinai (Reitter 1916a).

General distribution. Saudi Arabia, Yemen (Kaszab 1981).
Comments. The occurrence of this species in the Sinai is questionable (Andres 1931b, Gridelli 1953, Kaszab 1981).

6. *Adesmia (Oteroscelis) metallica metallica* (Klug, 1830)

Pimelia metallica Klug, 1830: No 23.
Adesmia faremontii Lucas, 1844: 264.
Referenced localities. The Sinai (Andres 1926, 1931b, Crotch 1872), Mazar (Boyd 1917 sub *A. faremonti*).
General distribution. Morocco up to Egypt (Koch 1940b).
Comments. The examination of several hundred specimens of the group of *Adesmia metallica* from the whole distribution area showed the distinctive criteria proposed by Koch (1940b) for *A. metallica metallica* and *A. metallica faremontii* are not constant. Beetles occur in the distribution area of *A. metallica* s. str. (sensu Koch) which do not differ from those from the area of *A. metallica faremontii* (sensu Koch). Besides, the nominate subspecies is to be found mainly in Libya under the forma *A. metallica faremontii*, and more rarely also in Algeria and Tunisia. This confirms the assertion of Gridelli (1930) who recognised both forms as synonymous. According to the collection site, the specimen reported by Boyd (1917) must be *A. metallica brozai* Ardoin, 1978.

7. *Adesmia (Adesmia) metallica syriaca* Baudi di Selve, 1874

Adesmia metallica var. *syriaca* Baudi di Selve, 1874: 95.
Referenced localities. Wadi Meghara (Alfieri 1976), the Sinai (Andres 1931b, Kaszab 1981).
General distribution. Israel, Jordan (Ardoin 1978), Iraq (Kaszab 1981).
Comments. It was probably confused also with *A. metallica brozai* Ardoin, 1978.

8. *Adesmia (Adesmia) montana acervata* (Klug, 1830)

Pimelia acervata Klug, 1830: No 26.
Referenced localities. Wadi Ain el Gederat, Wadi Isla (Alfieri 1976), the Sinai (Gridelli 1930).
General distribution. Algeria, Tunisia, Libya, Mediterranean Egypt (Koch 1949).
Comments. Only *A. montana montana* (Klug, 1830) has been surely identified from the Sinai. *A. montana acervata* differs from *A. montana* s. str. in the tubercles of the elytra which are smaller and somewhat more pointed in *A. montana acervata*. In some specimens of Lower Egypt, the differences from *A. montana* s. str. are very small.

9. *Adesmia (Adesmia) tenebrosa* Solier, 1835
Pimelia tenebrosa Solier, 1835b: 536.

Referenced localities. Djebel Watiyeh (Hart 1891, Peyerimhoff 1907 t. Hart).
General distribution. Iran (Solier 1835b).

Incorrect and Doubtful Records   109

Comments. *A. tenebrosa* is surely not present in the Sinai. It has probably been mistaken with the resembling *Adesmia montana montana* (Klug).

10. *Allophylax variolosus* (Olivier, 1811)

Opatrum variolosus Olivier, 1811: 497.
Comments. According to Reichard (1936), this species is to be found on the Iberian Peninsula, Morocco, Algeria and Egypt. It is missing in the catalogues of Alfieri (1976), Andres (1931b) and Koch (1935a). The information given by Innes Bey (1912) with regard to Wadi Genneh is questionable.

11. *Blaps laticollis* Solier, 1848

Blaps laticollis Solier, 1848: 324.
Referenced localities. Wadi Helal, Gebel El Arish (Alfieri 1976).
General distribution. Egypt (Solier 1848), the Sinai (Alfieri 1976), but not mentioned for the Sinai in Iwan et al. (2020).

12. *Cheirodes* (*Cheirodes*) *granulatus* (Laporte de Castelnau, 1840)

Anemia granulata Laporte de Castelnau, 1840: 218.
Referenced localities. Ain Musa, Berg Sinai (Walker 1871 sub *Anemia granulata*), the Sinai (Innes Bey 1912).
General distribution. Cape Verde Islands, Senegal up to Sudan (Ardoin 1971a), Saudi Arabia (Kaszab 1982).
Comments. This species had not been reported for the Sinai since Walker (1871) and Innes Bey (1912). The questionable specimens are destroyed. The distribution south of the Sahara mentioned by Ardoin (1971a) makes the presence of this species on the Sinai Peninsula appear doubtful.

13. *Clitobius ovatus* (Erichson, 1843)

Opatrum ovatum Erichson, 1843: 249.
Referenced localities. South of Ain Musa, at the foot of Djebel Hamam (Andres 1920).
General distribution. Namibia, Angola, Senegal, Cape Verde Islands, Morocco, Canary Islands, Sicily, Malta, Tunisia, Libya, Egypt (Grimm 1995), Turkey, Oriental region (Iwan et al. 2020).
Comment. The species was described from "Egypte". Seidlitz (1893) did not know it, also in Andres (1931b), Koch (1935a) and Purchart & Kamiński (2017) it is not mentioned for the Sinai.

14. *Crypticus subpunctatus* Walker, 1871

Crypticus subpunctatus Walker, 1871
Comments: This species has not been clearly determined. Walker (1871) describes the species without naming the collection site. According to Innes Bey

(1912) most of the insects described by Walker (1871) have been destroyed. On the basis of the remaining labels, he regards Wadi Feiran as the collection site of the questionable species.

15. *Erodius bilineatus* Olivier, 1791

Erodius bilineatus Olivier, 1791: 426.
Referenced localities. Magdaba (Alfieri 1920, 1976).
General distribution. *E. bilineatus bilineatus*: coastal dunes of Senegal and Mauritania (Ardoin 1971b). – *E. b. moccai* Koch, 1937: 327: Libya (Socna, Mizda, Fezzan (Hon) (Koch 1937). – *E. b. bungemensis* Koch, 1937: 328: Libya (Bu Ngem between Gheddahia and Hon) (Koch 1937). – *E. b. rohlfsi* Koch 1941: 258: Libya (Cyrenaica) (Koch 1941b).
Comments. Reports from Egypt need to be confirmed (Ardoin 1963). The report by Alfieri (1920, 1976) is probably based on an incorrect identification. Already Andres (1931b) noted that this species does not occur in Egypt.

16. *Erodius glabratus* Solier, 1834

Erodius glabratus Solier, 1834: 544.
Referenced localities. Wadi Mora (Andres 1931b), Wadi Karm Alam (Alfieri 1976).
General distribution. Saudi Arabia (Kaszab 1981), ? Egypt, ? the Sinai.
Comments. Described from Arabia (Solier 1834). Solier also mentions a specimen in the Gory collection originating from Egypt. The only specimen from Cairo we examined and that had been referred to by Kühnelt as *E. glabratus* (♀, NHMW, coll. Kühnelt) differs from the beetles of the Arabian type series (MNHNP) and from other specimens of the vicinity of Djidda (HJB, ML) by its bigger size and above all by its strongly punctured pronotum. These are probably two different species. The occurrence of *E. glabratus* outside of Arabia seems questionable.

17. *Erodius servillei* Solier, 1834

Erodius servillei Solier, 1834: 540.
Erodius scaber Solier, 1834: 542.
Referenced localities. Wadi Chamile (Heyden 1899 sub *E. scaber*), the Sinai (Andres 1931b, Gebien & Borchmann 1927, Holdhaus 1919, Reitter 1914a), Wadi Khamileh (Peyerimhoff 1907 t. Heyden sub *E. scaber*).
General distribution. Syria, Iraq, Iran, Arabia (Kaszab 1981), ? Sinai.
Comments. The species has been described as from Egypt although *E. servillei* auct. (=? Solier) does not occur there. As for the specimens reported from the Sinai, they would appear to be *E. puncticollis sinaiticus* Crotch (vgl. Koch 1940a).

18. *Erodius puncticollis puncticollis* Solier, 1834

Erodius puncticollis Solier, 1834: 556.
Erodius contractus auct. nec Kraatz, 1865: 31.

Erodius ecostatus Crotch, 1872: 267.
Referenced localities. Between Abu Diab and El Mishedi, between Wadi Mezara and Um Shusha (Alfieri 1920 sub *E. puncticollis contractus* Kraatz), Wadi Isla (Alfieri 1976).
General distribution. Egypt (Koch 1935a), ? Sinai.
Comments. It is not mentioned for the Sinai by Iwan et al. (2020). The specimens found must be *E. puncticollis sinaiticus*.

19. *Erodius octocostatus octocostatus* Peyerimhoff, 1907

Erodius octocostatus Peyerimhoff, 1907: 28.
Referenced localities. In the north of the Sinai Peninsula (Andres 1931b), the Sinai (Blair 1925, 1933, Kaszab 1981, Koch 1940a, Reitter 1914a).
General distribution. Iraq, SW Jordan, ? Arabia (Kaszab 1981), ? Sinai.
Comments. Peyerimhoff (1907) writes, *un peu au nord de la péninsule Sinaïtique* [...] *In deserto Sabaeo, loco dicto Hisme lectus*. It appears to be the northwestern Arabian region of Hisma crossed by the wadi Saba. The description of *Erodius octocostatus verruculiferus* Kaszab, 1982: 308 has been based on material collected in Saudi Arabia.

20. *Gonocephalum sericeum* (Baudi di Selve, 1875)

Opatrum sericeum Baudi di Selve 1875: 701.
Referenced localities. Wadi Genneh (Innes Bey 1912, Walker 1871).
Comments. Innes Bey (1912) names the *Hopatrum sericeum* Reiche from Wadi Genneh dealt of by Walker (1871), whereas Walker (1871) speaks of this species as *Hopatrum sericinum* Reiche from Cairo. According to Ferrer (1995a), *Opatrum sericeum* Baudi di Selve (not Reiche) is synonymous with *Gonocephalum patruele patruele*. Later, Ferrer (2010) revalidated *G. sericeum* (Baudi di Selve). Iwan et al. (2020) name Egypt, Libya, Israel, Saudi Arabia, furthermore the Afrotropis and the Nearctic Regions as distribution area. The record from Wadi Genneh needs confirmation.

21. *Gonocephalum strigosum* (Reiche, 1847)

Opatrum strigosum Reiche, 1847: 369.
Referenced localities. Mazar (Boyd 1917).
General distribution. Dahomey (Ferrer 1995a).
Comments. According to Ferrer (1995a), only one specimen is known from Dahomey (Benin) apart of the unique specimen without indication of collection site (Ardoin 1965, Ferrer 1995a). All other specimens determined as *G. strigosum* are wrongly named.

22. *Gonocephalum tomentosum* (Walker, 1871)

Hopatrum tomentosum Walker, 1871: 16.
Referenced localities. Wadi Gharandel (Gridelli 1953 t. Walker 1871, Innes Bey 1912), Peyerimhoff 1907 t. Walker 1871, Walker 1871).

General distribution. The Sinai (Type), ? Arabia (Kaszab 1982).
Comments. According to Innes Bey (1912) and Kaszab (1982) the type has been destroyed. The species cannot be identified. Even the genus belonging cannot be determined reliably according to the description.

23. *Imatismus (Curimosphena) sublinearis* **Walker, 1871**

Himatismus sublinearis Walker, 1871: 15.
Referenced localities. Wadi Gneh (Peyerimhoff 1907 t. Walker 1871), Wadi Genneh (Walker 1871).
General distribution. The Sinai (Walker 1871).
Comments. After Walker (1871), nobody seems to have examined these types. Peyerimhoff (1907) quotes the description, Andres (1931b) does not know the species and Alfieri (1976) and Koch (1935a) do not mention it. It appears to be *H. villosus* (Haag-Rutenberg, 1870).

24. *Imatismus (Curimosphena) variegatus* **(Fabricius, 1781)**

Tenebrio variegatus Fabricius, 1781: 323.
Referenced localities. El Arish, Magdaba (Alfieri 1920).
General distribution. Senegal, Bogos (Gebien 1937).
Comments. The data on *Himatismus variegatus* for the Sinai are based on determination mistakes. According to Alfieri (1976), it appears to be *H. villosus*.

25. *Lagria* spec.

Referenced locality. St Katherine Protectorate, V.-VI.2010 (Norfolk et al. 2012).
General distribution. Unknown as we do not know what species was recorded.
Comments. Unfortunately, this species was identified only to the genus level. We did not see any specimen of the genus from the Sinai. The genus *Lagria* has not yet been reported from the entire region (Israel, Lebanon, Jordan, the Arabian Peninsula, Libya and the African part of Egypt) (Iwan et al. 2020)

26. *Mesostena (Saxistena) cavatica* **(Andres, 1926)**

Mesostenopa cavatica Andres, 1926: 290.
Referenced localities. St Katherine Protectorate (Zalat *et al.* 2008).
General distribution. Egypt (Toura (Mokkatam), Wadi Digla near Cairo) (Andres 1926, 1931, Alfieri 1976, Kaszab 1982).
Comments. The cavernicolous species is so far known only from the vicinity of Cairo. Its occurrence in the Sinai Peninsula seems questionable.

27. *Mesostena (Mesostenopa) nabathaea* **(Peyerimhoff, 1907)**

Mesostenopa nabathaea Peyerimhoff, 1907: 31.
Referenced localities. The Sinai (Kaszab 1981 sub *M. nabathea*, Schawaller 1982 t. Kaszab 1981, Schuster 1928).

General distribution. Jordan (Peyerimhoff 1907).
Comments. The species was not described, as Kaszab (1981) reports, on the base of material collected from Ain El Hondra in the Sinai, but from Jordan (Petra). According to Andres (1931b) and Alfieri (1976) reports from the Sinai are incorrect.

28. *Mesostena (Mesostenopa) picea picea* (Kraatz, 1865)

Mesostenopa picea Kraatz, 1865: 180.
Referenced localities. Wadi Isla (Alfieri 1976), desert of Ga'a, Wadi Gergir, Wadi Budr (Andres 1920), Qantara (Boyd 1917 sub *Mesostenopa picea* Klug <!>), the Sinai (Andres 1931b, Crotch 1872, Gridelli 1938) garden of the Sinai Monastery (Heyden 1899), Monastery of St. Catherine (Peyerimhoff 1907 t. Heyden), Gulf of Aqaba, Wadi Feiran (Peyerimhoff 1907).
General distribution. The species colonises northern Africa with seven subspecies (Koch 1940c). According to Koch (1940c) *M. picea sinaitica* (Schatzmayr & Koch, 1934), and not *M. picea picea*, occurs in the Sinai.

29. *Microtelus asiaticus* Solier, 1838

Microtelus asiaticus Solier, 1838a: 10.
Referenced localities. Mount Sinai (Solier 1838a, Peyerimhoff 1907 t. Solier).
General distribution. Greece (Peloponnes, Crete), Turkey (region around Izmir) (Carl 1992a).
Comments. Described from Mount Sinai. This species was not found there since then.

30. *Oxycara aelaniticum* Peyerimhoff

Referenced localities. Tor (Bodenheimer & Theodor 1929).
Comments. Peyerimhoff did not describe any *Oxycara aelaniticum*. Bodenheimer & Theodor (1929) probably meant *Zophosis aelaniticum* Peyerimhoff (= *Z. nigroaenea* Deyrolle).

31. *Oxycara (Oxycara) laevigatum* (Reiche & Saulcy, 1857)

Melanchrus laevigatus Reiche & Saulcy, 1857: 192.
Referenced localities. Wadi Bel Maï (Heyden 1899).
General distribution. Israel, Jordan (Kaszab 1981), Syria (SMNS). According to Kaszab (1981) also in Oman.
Comments. The species recorded by Heyden is probably a species of the subgenus *Symphoxycara*. The examination of numerous *Oxycara* from Oman did not result in the discovery of any *O. laevigatum*. However, other species very similar to *O. laevigatum* could be found (Lillig 2009).

32. *Oxycara (Symphoxycara) subcostatum* (Guérin-Méneville, 1862)

Melancrus subcostatus Guérin-Méneville, 1862: 375.
Referenced localities. Wadi Isla, Wadi Um Mitla (Alfieri 1976), southern part of the Sinai Peninsula (Andres 1931b).

General distribution. The range of this species is limited to Yemen and some of the neighbouring islands.
Comments. The specimens from the Sinai must belong to *Oxycara peyerimhoffi* Lillig, 2001 (Lillig 2001).

33. *Pimelia (Pimelia) angulata nilotica* Sénac, 1884

Pimelia nilotica Sénac, 1884: 27.
Pimelia nilotica var. *aegyptiaca* Andres, 1929a: 15.
Referenced localities. Wadi Gharandel, Wadi Bel Maï, Wadi El Arish (Heyden 1899), Wadi Gharandel, Wadi El Arish (Peyerimhoff 1907 t. Heyden).
General distribution. Upper Egypt, Sudan (Koch 1940b, Leo 2020).
Comments. The beetles found in the Sinai must be the nominal form.

34. *Pimelia (Pimelia) bajula* Klug, 1830

Pimelia bajula Klug, 1830: No 8.
Referenced localities. The Sinai (Crotch 1872 sub *Pimelia bajula* Olivier <!>).
General distribution. Turkey ("Asia Minor"), Iraq, Syria, Israel (Koch 1941b), Cyprus (Grimm 1991), Jordan (Katbeh-Bader 1996), Egypt, Iran, Lebanon (Iwan et al. 2020).
Comments. It remains uncertain whether the species reported by Crotch is actually this one.

35. *Pimelia (Pimelia) derasa derasa* Klug, 1830

Pimelia derasa Klug, 1830: No 7.
Referenced localities. Wadi El Arish (Andres 1931b t. Heyden, Peyerimhoff 1907 t. Heyden), Wadi Bel Maï, Wadi El Arish (Heyden 1899).
General distribution. *P. derasa derasa*: littoral and sublittoral of Cyrenaica and Lower Egypt (Koch 1941b).
P. derasa barcana Koch, 1941: Gebel Barka (Koch 1941b).
Comments. The presence in the Sinai seems questionable.

36. *Pimelia (Pimelia) granulata tenuicornis* Solier, 1836

Pimelia tenuicornis Solier, 1836a : 99.
Referenced localities. Wadi Isla (Alfieri 1976 sub *Pimelia tenuicornis*).
General distribution. There are reports from Libya (Tripolitania), Egypt (Gridelli 1930) and the Sinai (Alfieri 1976). According to Kwieton (1977 and Iwan et al. 2020), it occurs only in Libya (Tripolitania).
Comments. Kwieton (1977) lists *P. tenuicornis* Solier as one of the eight subspecies of *P. granulata* SOLIER with *P. g. tenuicornis* as the most eastern subspecies from Tripolitania. A presence in the Sinai seems unlikely.

37. *Pimelia (Pimelia) sericea* Olivier, 1795

Pimelia sericea Olivier, 1795 (59): 8.
Pimelia asperata Klug, 1830: No 10.
Referenced localities. The Sinai (Crotch 1872 sub *Pimelia bajula* Olivier <!>).
General distribution. Lower Egypt (Alfieri 1976).
Comments. The occurrence of this species in the Sinai is questionable.

38. *Scaurus barbarus* Solier, 1838

Scaurus barbarus Solier, 1838b: 165.
Referenced localities. Mohammedia (Boyd 1917).
General distribution. Moroccan-Algerian border up to Cyrenaica (Peyerimhoff 1948b).
Comments. Surely not present in the Sinai.

39. *Scaurus tristis* Olivier, 1795

Scaurus tristis Olivier, 1795: 4.
Referenced localities. Ain Musa (Peyerimhoff 1907 t. Walker, Walker 1871).
General distribution. Western Mediterranean zone (Grimm 1985, Peyerimhoff 1948b).
Comments. Is not to be expected in the Sinai due to its western Mediterranean distribution.

40. *Scleron sulcatum saharense* Peyerimhoff, 1931

Scleron sulcatum saharense Peyerimhoff, 1931: 103.
Referenced localities. Wadi Isla (Alfieri 1976, Gridelli 1953 t. Koch 1935a, Koch 1935a).
General distribution. *S. sulcatum sulcatum* Baudi di Selve, 1876: 59: Arabia. – *S. sulcatum scotti* Gridelli, 1953: 58: Yemen. – *S. sulcatum lomii* Gridelli, 1953: 59: Arabia, tropical eastern Africa. – *S. sulcatum saharense*: central Sahara, the Sinai (Gridelli 1953).
Comments. The only specimen known from the Sinai is placed doubtfully by Koch (1935a) under *S. sulcatum saharense*.

41. *Tentyrina orbiculata orbiculata* (Fabricius, 1775)

Pimelia orbiculata Fabricius 1775: 253.
Referenced localities. Wadi Isla, Wadi Mezara, Um Shusha (Alfieri 1976), Wadi Amarah (Peyerimhoff 1907).
Comments. It appears to be *T. orbiculata subsulcata* (Reiche & Saulcy, 1857). According to Koch (1940c) *T. orbiculata orbiculata* occurs only in Lower Egypt.

42. *Tentyrina longicollis* (Lucas, 1855)

Tentyria longicollis Lucas, 1855: 291.
Referenced localities. The Sinai (Pierre 1963).
General distribution. Morocco (Kocher 1958), Algeria, Tunisia, Libya (Koch 1940c).
Comments. According to Pierre (1963), the distribution range of this species is the Sinai, Mesopotamia, Arabia, Tibesti and Fezzan. This information contradicts other authors (Iwan et al. 2020, Koch 1940a, Kocher 1958) and Lucas (1855), who described the species from Algeria.

43. *Zophosis (Septentriophosis) bicarinata quadricostata* Solier, 1834

Zophosis quadricostata Solier, 1834: 628.
Referenced localities. South of Ain Musa up to Wadi Werdan, Wadi Hebran, desert of Ga'a, Wadi Chaschiba (Andres 1920 sub *Z. quadricostata*), Wadi Amara, Wadi Nasb (Peyerimhoff 1907 sub *Z. quadricostata* t. Walker, Peyerimhoff 1933 sub *Z. quadricostata*, Walker 1871 sub *Z. quadricostata*), Aïn el-Houdrâ, Wadi el Aïn (Peyerimhoff 1907 sub *Z. quadricostata*), the Sinai (Kaszab 1981).
General distribution. Along the Red Sea from Jemini near Aqaba (Jordan) up to Yemen and Eritrea (Penrith 1982a).
Comments. *Z. bicarinata quadricostata* has been described on the basis of material collected in Egypt without more precise specification of the collection site. The published data suggest that the specimens from the Sinai are *Z. bicarinata* cf. *ghilianii*.

VIII Species Expected to Be Found on the Sinai Peninsula (In Alphabetical Order)

1. *Adesmia (Adesmia) montana parallela* **Miller, 1861**

Adesmia parallela Miller, 1861: 171.
Nearest known locality. Surroundings of Suez, leg. L. Vaillant 1864 (2) (MNHNP).
Referenced localities. Pharaoh's Baths (= Hammam Fir'aoun), Wadi Gharandel (Peyerimhoff 1907 t. Walker, Walker 1871).
General distribution. Arabian desert of Egypt, northern Sudan (Koch 1949).
Comments. A confirmation for the Sinai is still missing, even if the evidence from Suez (BM, MNHNP) makes the occurrence in the north-west of the Peninsula plausible.

2. *Cheirodes (Pseudanemia) submetallicus* **(Raffray, 1873)**

Anemia? *submetallica* Raffray, 1873: 377.
General distribution. North Africa north of the Sahara, from Morocco to Egypt, Sudan, Jericho (Ardoin 1971a), Jordan, Saudi Arabia (Iwan & Löbl 2020).
Comments. On the basis of its distribution, an occurrence on the Sinai Peninsula is to be expected.

3. *Cyptus aegyptiacus* **(Mulsant & Rey, 1859)**

Caedius aegyptiacus Mulsant & Rey, 1859: 136.
General distribution. Egypt, Sudan, Niger (Pierre 1972 sub *Neocaedius aegyptiacus*), Afghanistan, Israel (Iwan & Löbl 2020).
Comments. *C. aegyptiacus* was found in Port Said (Pierre 1972). An occurrence on the eastern bank of the Suez Canal is possible.

4. *Erodius (Eodirosis) quadrilineatus* **Kraatz, 1865**

Erodius quadrilineatus Kraatz, 1865: 31.
General distribution. Egypt, Israel (the Negev) (Lillig 1994).
Comments. This species is to be expected in the north of the Sinai.

5. *Gonocephalum (Gonocephalum) controversum* **Gridelli, 1948**

Gonocephalum controversum Gridelli, 1948: 11.
General distribution. Syria, Egypt, Sudan, Eritrea, Chad (Ferrer 2000b), Palestine (Español 1973), Uganda (Gridelli 1948), Bahrain, Jordan, Saudi Arabia, Yemen (Iwan & Löbl 2020).
Comments. The species could also be found in the Sinai.

6. *Gonocephalum (Gonocephalum) costatum rugulosum* **(Küster, 1849)**

Opatrum rugulosum Küster, 1849: 55.
General distribution. Turkey, Iran, Israel (Ferrer 1995a), Armenia, Greece (Lesbos), Cyprus, Syria (Iwan & Löbl 2020).
Comments. *G. costatum rugulosum* was found in Eilat, Israel (Ferrer 1995a). It could also occur on the Sinai Peninsula.

7. *Mesomorphus longulus* **(Reiche & Saulcy, 1857)**

Crypticus longulus Reiche & Saulcy, 1857: 263.
General distribution. Greece (Island of Kos), Cyprus, Turkey, Israel (Ferrer 2000a), Iran, Jordan, Lebanon, Syria (Iwan & Löbl 2020).
Comments. In all Israel, also Ze'elim near the border to the Sinai (ML). To be expected in the north/east of the Sinai.

8. *Micipsa batesi* **Haag-Rutenberg, 1875**

Micipsa batesi Haag-Rutenberg, 1875b: 92.
General distribution. "Syria" (Haag-Rutenberg 1875b), the Negev (Bytinski-Salz 1955a).
Comments. An occurrence in the northern part of the Sinai seems possible.

9. *Microtelus lethierryi* **Reiche, 1860**

Microtelus lethierryi Reiche, 1860: 334.
Microtelus torretassoi Koch, 1934: 24.
General distribution. In the distribution range of the species, Carl (1992a: 82) includes the northern part of the Sinai in his generalised map, although he does not name any collection site on the Peninsula. To be considered as sure are findings from Corsica (not in Iwan & Löbl 2020), Algeria, Tunisia, the Egyptian Mediterranean coast and Israel (Carl 1992a), which makes an occurrence in the Sinai plausible. Besides, *M. lethierryi* has been reported from Italy (Island of Lampedusa) (Peyerimhoff 1931), Sicily (Gridelli 1930), Morocco (Iwan & Löbl 2020, Kocher 1958, Peyerimhoff 1944) and the former Spanish Sahara (Español 1943).

10. *Phaleria acuminata syriaca* **Reitter, 1916**

Phaleria syriaca Reitter, 1916b: 7.
General distribution. Mediterranean coast of Egypt, Israel and Lebanon (Canzoneri 1968).
Comments. The distribution gap on the Mediterranean coast of the Sinai is probably artificial.

11. *Philhammus (Philhammus) aharonii* **(Reitter, 1910)**

Psilachnopus aharonii Reitter, 1910: 197.
General distribution. Egypt, Jordan, Israel (Kaszab 1967), Sudan (Lillig & Bremer 2002), Morocco, Saudi Arabia, Iran (Iwan & Löbl 2020).
Comments. To be expected in the north of the Sinai.

12. *Pimelia (Pimelia) derasa derasoides* **Schuster, 1922**

Pimelia (Pimelia) derasoides Schuster, 1922: 19.
General distribution. Jordan (Katbeh-Bader 1996), Jordan valley (Koch 1941b), Israel (Negev, Ze'elim, ML), Syria (Iwan & Löbl 2020).
Comments. The specimens published under *P. derasa* Klug, 1830: No 7 belong probably to *P. derasoides*.

13. *Pimelia (Pimelia) interpunctata* **Klug, 1830**

Pimelia interpunctata Klug, 1830: No 2.
Doderoella cyrenaica Schuster, 1926: 134.
Doderoella andreinii Gridelli, 1929a: 10.
General distribution. Libya (Tripolitania) (Koch 1937), Egypt (Alexandria) (Alfieri 1976), Israel (Negev, Ze'elim) (ML).
Comments. One can assume that with further searching in the northern part of the Sinai the species could be found there. The examination of the type of *P. interpunctata cyrenaica* (GF), a specimen determined by Koch as *P. interpunctata andreinii* from Tauorga, north-east of Tripolitania, IV.<19>38, G. Frey (GF) as well as of several specimens of Egyptian *P. interpunctata* (HJB, GF) shows that the forms listed as subspecies are only varieties (Lillig & Pavlíček 2003). This was already indicated by Koch (1937). Formerly, *P. interpunctata* was listed under the monotype genus *Doderoella*. Kwieton (1982) unifies this genus with *Pimelia* s. str.

14. *Pseudoseriscius griseovestis* **(Fairmaire, 1879)**

Crypticus griseovestis Fairmaire, 1879a: 192.
General distribution. Algeria, Tunisia, Libya (Cyrenaica), Egypt (Español 1949), Cyprus, Jordan, Kuwait (Iwan & Löbl 2020).

Comments. Español (1949) names two collection sites at the border of the Sinai, Ismailia and Port Said, so that an occurrence on the Peninsula appears possible.

15. *Sclerum armatum* (Waltl, 1835)

Opatrum armatum Waltl, 1835: 72.
General distribution. Spain, Sicily, Algeria, Tunisia (Gridelli 1930), Libya (Koch 1937), Egypt, Jordan (Katbeh-Bader 1996), France, Spain, Syria, Morocco, Canary Islands (Iwan & Löbl 2020).
Comments. On the basis of its distribution, the species could also be found in the Sinai.

16. *Sclerum multistriatum* (Forskål, 1775)

Silpha multistriata Forskål, 1775: 77.
General distribution. Southern Russia, Greece, Crete, "Syria", Turkey, Israel, Lower Egypt, Libya (Cyrenaica), Sicily (Gridelli 1930), Malta, Syria (Iwan & Löbl 2020).
Comments. Grimm (1981) quotes different opinions regarding the distribution of this species. It is to be expected in the Sinai.

17. *Tenebrio syriacus* Zoufal, 1892

Tenebrio syriacus Zoufal, 1892: 284.
General distribution. Lower Egypt (Alfieri 1976, Koch 1935a), "Syria" (Zoufal 1892), Turkey (Iwan & Löbl 2020).
Comments. The species could also be found in the Sinai.

Bibliography

Alfieri, A. (1920) Contribution à la faune entomologique du Nord-Est du Sinaï. – Bulletin de la Société entomologique d'Égypte 6: 40–55; Cairo.
Alfieri, A. (1976) The Coleoptera of Egypt. – Mémoires de la Société entomologique d'Égypte 5: 1–359; Cairo.
Aliquó, V. & Soldati, F. (2010) Coleotteri Tenebironidi di Sicilia (Col. Tenebrionidae). Edizioni Danaus, 175; Palermo.
Allard, E. (1865) Quatre nouvelles espèces du genre *Erodius*. – Annales de la Société entomologique de France 4 (4. série): 387–388; Paris.
Allard, E. (1882) Descriptions de nouvelles espèces de Coléoptères. – Bulletin entomologique 2 (6e série): lxxxvi–lxxxvii; Paris.
Allard, E. (1883) Mélanges entomologiques. – Annales de la Société entomologique de Belgique 28: 7–49; Bruxelles.
Allard, E. (1885) Classification des Adesmides et Mégagénides (Lacord.). – Annales de la Société entomologique de France 5 (6. série): 155–208; Paris.
Ancey, C. M. F. (1881) Descriptions de coléoptères nouveaux d'Aden. – Le Naturaliste 3: 397; Paris.
Andres, A. (1920) Zoologische Ergebnisse zweier in den Jahren 1902 und 1904 durch die Sinaihalbinsel unternommener botanischer Studienreisen. II. Teil. I. Tenebrionidae (Coleoptera). – Entomologische Blätter 16: 61–74; Krefeld.
Andres, A. (1921) Notes sur divers Ténébrionides d'Égypte (Col.). – Bulletin de la Société entomologique d'Égypte 6: 67–75; Cairo.
Andres, A. (1922) Zoologische Ergebnisse zweier in den Jahren 1902 und 1904 durch die Sinaihalbinsel unternommener botanischer Studienreisen. Nachtrag. Tenebrionidae. – Entomologische Blätter 18: 26–27; Krefeld.
Andres, A. (1926) Notes et description d'un Tentyrinide d'Égypte: *Meostenopa cavatica* sp. n. (Coléop.-Ténébrionide). – Bulletin de la Société Royale Entomologique d'Égypte 9 [1925]: 287–290; Cairo.
Andres, A. (1927a) Sur l'*Adesmia metallica* Klug (Col.-Ténébrion.) et ses variétés. – Bulletin de la Société Royale Entomologique d'Égypte 10 [1926]: 257–262; Cairo.
Andres, A. (1927b) Sur divers Ténébrionides d'Égypte. – Bulletin de la Société Royale Entomologique d'Égypte 11: 52–55; Cairo.
Andres, A. (1928) Révision des espèces égyptiennes du genre *Ocnera* (Coleop. - Tenebrionidae). – Bulletin de la Société Royale Entomologique d'Égypte 12: 36–39; Cairo.
Andres, A. (1929a) Note sur *Pimelia angulata* Fabr. et Espèces voisines et Description d'une Variété nouvelle. – Bulletin de la Société Royale Entomologique d'Égypte 13: 126–148; Cairo.
Andres, A. (1929b) Note synonymique sur les *Thriptera* égyptiennes (Coléoptère: Ténébrionidae). – Bulletin de la Société Royale Entomologique d'Égypte 13: 10–24; Cairo.
Andres, A. (1931a) A new Tenebrionid from Egypt: *Mitotagenia Prieneri*. – Bulletin de la Société Royale Entomologique d'Égypte 15: 56–57; Cairo.
Andres, A. (1931b) Catalogue of the Egyptian Tenebrionidae. – Bulletin de la Société Royale Entomologique d'Égypte 15: 74–125; Cairo.
Anonymous (2002) A Consolidated Replacement Text for the Nomination of the St. Catherine area, South Sinaï, Egypt, as a mixed world cultural and natural heritage site, 63 pp. plus 10 pls.; whc.unesco.org/ (pdf-document, last accessed on 28.12.2021).

Ardoin, P. (1963) Récoltes de M. A. Villiers dans les dunes côtières du Sénégal (1961). Coléoptères Tenebrionidae. – Bulletin de l'Institut fondamental d'Afrique noire 25, sér. A (2): 372–388; Dakar.

Ardoin, P. (1965) Contribution à l'étude des Ténébrionides africains (Col.). – Bulletin de la Société entomologique de France 70: 73–85; Paris.

Ardoin, P. (1971a) Contribution à l'étude des espèces africaines et malgaches du genre *Anemia* Laporte (Col. Tenebrionidae). – Annales de la Société entomologique de France (N.S.) 7 (2): 357–422; Paris.

Ardoin, P. (1971b) Contribution à l'étude biologique du Sénégal septentrional. VIII. Coléoptères Tenebrionidae. – Bulletin de l'Institut fondamental d'Afrique noire 28, sér. A (1): 102–124; Dakar.

Ardoin, P. (1972a) Liste des espèces de Tenebrionidae (Coleoptera) récoltées au Soudan par les expéditions finlandaises (1962-1964). – Commentationes Biologicae 49: 1–20; Helsinki.

Ardoin, P. (1972b) Tenebrionidae (Coleoptera) récoltés dans les environs de Rosso, Mauritanie par M. J.-L. Amiet. – Annales de la Faculté des Sciences de l'Université fédérale du Cameroun 10: 85–105; Yaoundé.

Ardoin, P. (1973) Contribution à l'étude des Tenebrionidae (Coleoptera) de Sardaigne. – Annales de la Société entomologique de France (N.S.) 9 (2): 257–307; Paris.

Ardoin, P. (1974) Recherches sur le littoral de la Somalie. La plage et la dune de Sar Uanle. 1. Tenebrionidae (Coleoptera). – Monitore Zoologico Italiano (N. S.) Supplemento 5 (12): 133–145; Firenze.

Ardoin, P. (1978) Contribution à l'étude des *Adesmia* (Coleoptera: Tenebrionidae) du Sinaï, d'Israël et de Jordanie. – Nouvelle Revue d'Entomologie 8 (3): 295–313; Toulouse.

Ayal, Y. & Merkl, O. (1994) Spatial and temporal distribution of tenebrionid species (Coleoptera) in the Negev Highlands. – Journal of Arid Environments, 27 (4): 347–361; Oxford.

Ayyad, M. A. & S. I. Ghabbour (1986) Hot Deserts of Egypt and the Sudan. – In: Avenari, M., Noir-Meir, I. & Goodall, D. W. (eds.): Ecosystems of the world. Volume 12B. Hot deserts and arid shrublands: 149–202; Amsterdam.

Baha El Din, S. (1999) Directory of Important Bird Areas in Egypt. Bird Life International, 113; Cairo.

Baldi, M., Amin, D., Al Zayed, I. S. & Dalu, G. (2020) Climatology and Dynamical Evolution of Extreme Rainfall Events in the Sinai Peninsula-Egypt. – Sustainability 12: 61–86; Basel. DOI: 10.3390/su12156186

Bates, F. (1879) Characters of the new genera and species of Heteromera collected by Dr. Stoliczka during the Forsyth Expedition to Kashgar in 1873-74. – Cistula Entomologica 2 [1875-1882]: 467–484; London.

Baudi di Selve, F. (1874) Catalogo dei Tenebrioniti della fauna europea e circummediterranea appartenenti alle collezioni del Museo Civico di Genova. Parte prima. – Annali del Museo Civico di Storia naturale di Genova 6: 89–115; Genova.

Baudi di Selve, F. (1875) Catalogo dei Tenebrioniti della fauna europea e circummediterranea appartenenti alle collezioni del Museo Civico di Genova. Parte seconda. – Annali del Museo Civico di Storia naturale di Genova 7: 684–703; Genova.

Baudi di Selve, F. (1876a) Coleotteri Tenebrioniti delle collezioni italiane. – Bollettino della Società Entomologica Italiana 7: 209–237; Firenze.

Baudi di Selve, F. (1876b) Europaeae et circummediterraneae Faunae Tenebrionidum specierum, quae Comes Dejean in suo Catalogo, editio 3a, consignavit, ex ejusdem collectione in R. Taurinensi Musaeo asservata, cum auctorum hodierne recepta denominatione collatio. – Deutsche Entomologische Zeitschrift 20 (1): 1–74; Berlin.

Baudi di Selve, F. (1881) Heteromerum species ex Aegypto, Syria et Arabia. – Deutsche Entomologische Zeitschrift 25 (2): 273–296; Berlin.

Baudi di Selve, F. (1883) Descrizione d'una specie di *Zophosis*. – Annali del Museo civico di Storia naturale Giacoma Doria 18: 760–761; Genova.
Baudi di Selve, F. (1894) Viaggio del Dr. E. Festa in Palestina, nel Libano e regioni vicine. – Bollettino di Musei di Zoologia ed Anatomia comparata della R. Università di Torino 9 (173): 1–13; Torino.z7
Billberg, G. J. (1820) Enumeratio insectorum in museo Gust. Joh. Billberg. Typis Gadelianis, [4]+138; Stockholm.
Blair, K. G. (1914) On the Fabrician Types of Tenebrionidae (Coleoptera) in the Banks Collection. – Annals and Magazine of Natural History, including Zoology, Botany and Geology (8) 13: 482–490; London.
Blair, K. G. (1925) A list of the heteromerous Coleoptera collected mainly during the war, by various members of H. M. Forces in Mesopotamia. – Journal of the Bombay Natural History Society 30: 428–440; Bombay.
Blair, K. G. (1931) A collection of Coleoptera from the Rub'al Khali, Southern Arabia, with descriptions of some new species and varieties. – The Entomologist's Monthly Magazine 67: 269–271; London.
Blair, K. G. (1933) Tenebrionidae from central Arabia collected by Major H. St. J. B. Philby. – Entomologist's Monthly Magazine 69: 4–7; London.
Blair, K. G. (1934) On Coleoptera from the Ahaggar Mountains. – Annals and Magazine of Natural History, including Zoology, Botany and Geology (10) 13: 181–185; London.
Blair, K. G. (1935) Some synonymic notes in the family Tenebrionidae (Col.). – Entomologist's Monthly Magazine 71: 102–104; London.
Blanchard, C. E. (1845) Histoire des insectes: traitant de leurs moeurs et de leurs métamorphoses en général et comprenant une nouvelle classification fondée sur leurs rapports naturels. In: Traité complet d'histoire naturelle. Histoiredes Insectes. Tome second. Didot, 524 pp; Paris.
Bodenheimer, F. S. (1937) Prodromus Faunae Palaestinae. – Mémoires de l'Institut d'Égypte 32: 1–286; Cairo.
Bodenheimer, F. S. & Theodor, O. (1929) Ergebnisse der Sinai-Expedition 1927 der Hebräischen Universität, Jerusalem. Leipzig, 143 pp., 14 pls.
Boehm, R. (1908) Les Adesmides des déserts égyptiens. – Bulletin de la Société Royale Entomologique d'Égypte 1908: 151–165; Cairo.
Boehm, R. (1909) Les Zophosini et Erodiini de l'Égypte. – Bulletin de la Société Royale Entomologique d'Égypte 1909: 108–127; Cairo.
Borkin, L. J., Goncharov, A. I. & Litvinchuk, S. N. (2016) The Egyptian toad, *Sclerophrys regularis* (Reuss, 1833) at Sharm El-Sheikh, with comments on amphibians of the Sinai Peninsula. – Russian Journal of Herpetology 23 (4): 283–292; Moscow. DOI: 10.30906/1026-2296-2016-23-4-283-292
Botosaneanu, L. (1999) Caddisflies (Trichoptera). – What can they tell about biogeography of the Levant? – Israel Journal of Zoology 45: 75–78, Jerusalem.
Boyd, A. W. (1917) Six Month' collecting between Ismailia and El Arish. – Bulletin de la Société Entomologique d'Égypte 5: 98–119; Cairo.
Brullé, G. A. (1832) Expédition scientifique de Morée. Tome 3, Partie insectes; Paris, Levrault, t. 3, pt. 1, sect. 2, p. 64–395.
Bunalski, M. & Śmiełowski, J. (2018) Tenebrionidae (Coleoptera) of Nubia: Annotated Catalogue of Species, with New Data from the Bayada Desert. – The Coleopterists Bulletin 72 (3): 465–470; Brookings. DOI: 10.1649/0010-065X-72.3.465
Bytinski-Salz, H. (1955a) Collecting Tenebrionid Beetles with Poisoned Bait. – Entomologist's Record 67: 17–19; Surrey.
Bytinski-Salz, H. (1955b) *Prionotheca coronata* Ol. *freyi* ssp. n. – Entomologische Arbeiten aus dem Museum Georg Frey 6 (1): 313–316; Tutzing.

Canzoneri, S. (1968) Materiali per una monografia delle *Phaleria* del sottogenere *Phaleria* Latr. XX Contributo alla conescenza dei Coleoptera Tenebrionidae. – Memorie della Società entomolgica Italiana 47: 117–167; Genova.

Carl, M. (1990) Übersicht über die irakischen *Adesmia*-Arten und Neubeschreibung von *Adesmia migrogranulata* sp. n. (Coleoptera, Tenebrionidae). – Mitteilungen der Münchner Entomologischen Gesellschaft 80: 71–83; München.

Carl, M. (1991a) Die Gattungen *Adelostomoides* gen. n. und *Adelostoma* Dup. aus Mesopotamien (Coleoptera: Tenebrionidae). – Nachrichtenblatt der bayerischen Entomologen 40 (1): 23–27; München.

Carl, M. (1991b) Beitrag zur Schwarzkäferfauna Tadschikistans (UdSSR). – Spixiana 14 (3): 275–279; München.

Carl, M. (1992a) A revision of the genus *Microtelus* Solier, 1838 (Coleoptera: Tenebrionidae). – Zoology of the Middle East 6: 69–87; Heidelberg.

Carl, M. (1992b) Neue und bemerkenswerte Tenebrionidae (Coleoptera) aus dem Vorderen Orient. – Linzer biologische Beiträge 24/1: 331–337; Linz.

Carl, M. (1994) The Taxonomy, Biogeography and Phylogeny of the Genus *Prochoma* in the Middle East and South-East Asia (Coleoptera: Tenebrionidae). – Fauna of Saudi Arabia 14: 77–96; Basel.

Castro Tovar, A. (2020) Nuevas sinonimias en el género *Cheirodes* Gené, 1839 (Coleoptera: Tenebrionidae: Melanimini). – Boletín de la Sociedad Andaluza de Entomología (SAE) 30: 55–68; Sevilla.

Chevrolat, A. (1877) Coléoptères nouveaux de la famille des Ténébrionides. – Petites Nouvelles Entomologiques 2: 113–114; Paris.

Chikatunov, V., Lillig, M., Pavlíček, T., Blaustein, L. & Nevo, E. (1997) Biodiversity of insects at a microsite, „Evolution Canyon", Nahal Oren, Mt. Carmel, Israel. Coleoptera: Tenebrionidae. – Journal of Arid Environments 37: 367–377; Cambridge. DOI: 10.1006/jare.1997.0278

Chobaut, A. (1897) Description de deux espèces nouvelles de Coléoptères du Mzab (Sud-algérien). – Bulletin de la Société entomologique de France 1897: 204–206; Paris.

Cloudsley-Thompson, J. L. (1965) On the function of the sub-elytral cavity in desert Tenebrionidae (Col.). – Entomolgist's Monthly Magazine 100: 147–151.

Cloudsley-Thompson, J. L. (1978) Form and Function in Animals (Patterns of Progress: Zoology series 10). Meadowfield Press Ltd., 81 pp; Shildon.

Condamine, F. L., Soldati, L., Rasplus, J.-Y. & Kergoat, G. L. (2011) New insights on systematics and phylogenetics of Mediterranean *Blaps* species (Coleoptera: Tenebrionidae: Blaptini), assessed through morphology and dense taxon sampling. – Systematic Entomology 38 (2): 340–361; London. DOI: 10.1111/j.1365-3113.2010.00567.x

Constantinou, C. & Cloudsley-Thompson, J. L. (1983) On the status and distribution of *Trachyderma hispida* and *T. philistina* (Coleoptera: Tenebrionidae). – Journal of Arid Environments 6: 129–133; London. DOI: 10.1016/S0140-1963(18)31525-8

Crotch, G. R. (1872) Part 2. List of the Coleoptera found during the progress of the survey. – In: Wilson, C. W. & Palmer, H. S. (eds.): Ordnance Survey of the Peninsula of Sinai. Ordnance Survey Office, pp. 263–268; Southampton.

Crovetti, A. (1970) Note eco-etologiche sulla entomofauna primaverile dello "Uádi Caam" (Tripolitania). – Studi Sassaresi: Sezione III, Annali della Facoltà di agraria dell'Università di Sassari 18: 270–381; Sassari.

Crowson, R. A. (1981) The Biology of the Coleoptera. Academic Press, 802 pp; London. DOI: 10.1016/C2013-0-07304-5

Danin, A. (1983) Desert vegetation of Israel and Sinai. – Cana Publication House, 148 pp; Jerusalem.

Dejean, P. F. M. A. (1834) Catalogue des Coléoptères de la collection de M. le Comte Dejean. 2ème édition. 3ème livraison. Méquignon-Marvis Père et Fils, pp. 177–256; Paris.

Desbrochers des Loges, J. (1881) Insectes Coléoptères du nord de l'Afrique nouveaux ou peu connus. 1er Mémoire. Ténébrionides. – Bulletin de l'Académie d'Hippone 16: 51–167; Bone.
Deyrolle, A. (1867) Monographie de la tribu *Zophosites*. – Annales de la Société entomologique de France 7 (4. série): 73–248; Paris.
Duponchel, F. (1827) Description d'un nouveau genre d'insectes de l'ordre des Coléoptères (section des Hétéromères), trouvé dans l'île de Léon en 1824, et liste des insectes du même ordre recueillis dans cette île. – Annales de la Société Linnéenne de Paris 6: 338–343; Paris.
Eddy, F. W. & Wendorf, F. (1999) An Archaeological Investigation of the Central Sinai, Egypt. – American Research Centre in Egypt, Inc. and the University Press of Colorado, 340 pp; Boulder.
Erichson, W. F. (1841) Über die Insekten von Algier, mit besonderer Rücksicht auf ihre geographische Verbreitung. – In: Wagner, M. (ed.): Reise in der Regentschaft Algier in den Jahren 1836, 1837 und 1838. Dritter Band nebst einem naturhistorischen Anhang und einem Kupferatlas. Verlag von Leopold Voss. 296 pp., Leipzig.
Erichson, W. F. (1843) Beitrag zur Insecten-Fauna von Angola, in besonderer Beziehung zur geographischen Verbreitung der Insecten in Afrika. – Archiv für Naturgeschichte 9 (1): 199–267; Berlin.
Eschscholtz, J. F. (1829) Zoologischer Atlas enthaltend Abbildungen und Beschreibungen neuer Thierarten während des Flottcapitains von Kotzebue zweiter Reise um die Welt auf der Russisch-Kaiserlichen Kriegesschlupp Predpriaetië in den Jahren 1823-1826. Drittes Heft. G. Reimer, 18 pp, pls xii–xv; Berlin.
Eschscholtz, J. F. (1831) Zoologischer Atlas, enthaltend Abbildungen und Beschreibungen neuer Thierarten, während des Flottscapitains V. Kotzebue zweiter Reise um die Welt, auf der Russisch-Kaiserlichen Kriegsschlupp. Predpriaetië in den Jahren 1823-1826. Viertes Heft. – G. Reimer, 19 pp, pls xvi–xx, Berlin.
Español, F. (1943) Misión Científica E. Morales Agacino, Ch. Rungs y B. Zolotarevsky a Ifni y Sáhara Español. Tenebrionidae (Col.). 1a Parte. – EOS: Revista Española de Entomología 19: 119–148; Madrid.
Español, F. (1949) Contribución al conociemento de los *Crypticus* del grupo del *pruinosus*: el subgenero *Pseudoseriscius* Españ. (Col. Tenebrionidae). – EOS: Revista Española de Entomología 25: 199–239; Madrid.
Español, F. (1950) Los *Crypticus* s.str. de la Peninsula Ibérica y del norte de Africa (Col. Tenebrionidae). – EOS: Revista Española de Entomología 26: 115–156; Madrid.
Español, F. (1952) Más Tenebrionidos del Sáhara Español (Col.). – EOS: Revista Española de Entomología 28 (1): 71–76; Madrid.
Español, F. (1967) Misión entomológica Hakan Lindberg y M. Meinander a Marruecos. Col. tenebriónidos. – Publicaciones del Instituto de Biologia Aplicada 42: 17–47; Barcelona.
Español, F. (1973) Coleópteros Tenebriónidos recogidos por J. Mateu en el macizo del Ennedi y en el norte Tchad. – Bulletin de l'Institut Fondamental d'Afrique Noire 35, série A (2): 303–330; Dakar.
Fabricius, J. C. (1775) Systema entomologicae, sistens insectorum, classes, ordines, genera, species, adiectis synonymis, locis, descriptionibus, observationibus. Libraria Kortii, Flensburg, 832 pp.
Fabricius, J. C. (1781) Species insectorum, exhibens eorum differentias specificas, synonyma auctorum, loca natalia, metamorphosis, adiecitis observastionibus, descriptionibus.: Tom 1: viii + 552 pp., – Carol Ernest Bohnii: Hamburgi et Kilonii.
Fabricius, J. C. (1787) Mantissa insectorum, sistens eorum species nuper detectas adiectis characteribus genericis, differentiis specificis, emendationibus, observationibus. Volume 1: 348 pp. Hafniae.
Fairmaire, L. (1849) Essai sur les Coléoptères de la Polynésie (suite). – Revue et Magazine de Zoologie Pure et Appliquée 2e série 1: 445–460; Paris.

Fairmaire, L. (1868): Essai sur les Coléoptères de Barbarie. Sixième partie. – Annales de la Société Entomologique de France (4) 8: 471–502; Paris.
Fairmaire, L. (1869) Notes sur les Coléoptères recueillis par Charles Coquerel à Madagascar et sur les côtes d'Afrique. 1er partie. – Annales de la Société entomologique de France 8 (4): 753–820; Paris.
Fairmaire, L. (1870) Essai sur les Coléoptères de Barbarie. Septième partie. – Annales de la Société entomologique de France (4) 10: 369–404; Paris.
Fairmaire, L. (1875) Diagnose de quelques Coléoptères nouveaux d'Algérie. – Petites nouvelles entomologiques 1: 495–496; Paris.
Fairmaire, L. (1879a) Coléoptères du nord de l'Afrique. – Revue et Magazine de Zoologie 3, Série 7: 178–218; Paris.
Fairmaire, L. (1879b) Diagnose de Coléoptères du Nord de l'Afrique. – Le Naturaliste (2) 1: 4; Paris.
Fairmaire, L. (1880) Descriptions de Coléoptères nouveaux du nord de l'Afrique. 3e Partie. – Annales de la Société entomologique de France 10 (5. série): 5–32; Paris.
Fairmaire, L. (1886) Note sur les Coléoptères recueillis par M. Laligant à Obock. – Annales de la Société entomologique de France 5 (6. série): 435–462; Paris.
Fairmaire, L. (1892) Coléoptères d'Obock. Troisième partie. – Revue d'Entomologie 1892: 77–127; Paris.
Fairmaire, L. (1894) Descriptions de Coléoptères d'Algérie. – Annales de la Société Entomologique de Belgique 38: 310–313; Bruxelles.
Fairmaire, L. (1897) Description de trois Coléoptères hétéromères d'Algérie et d'Égypte. – Bulletin de la Société entomologique de France 1897: 120–121; Paris.
Faldermann, F. (1837) Coleoptera trans-caucasica. II. Heteromera. – Fauna entomologica transcaucasica. Pars II. – Nouveaux Mémoires de la Société impériale des Naturalistes de Moscou 5: 3–433, pls. 1–16; Moscow.
Fattorini, S. (2015) On the concept of chorotype. – Journal of Biogeography 42 (11): 2246–2251. DOI: 10.1111/jbi.12589
Ferrer, J. (1993) Essai de révision des espèces africaines et européennes appartenant au genre *Gonocephalum* Solier (Coleoptera, Tenebrionidae). – Atti del Museo Civico di Storia naturale di Trieste 45: 59–150; Trieste.
Ferrer, J. (1995a) Essai de révision des espèces africaines et européennes appartenant au genre *Gonocephalum* Solier (Coleoptera, Tenebrionidae). Deuxième partie. – Atti del Museo Civico di Storia naturale di Trieste 46: 1–75; Trieste.
Ferrer, J. (1995b) Contribution to the knowledge of the Tenebrionidae of Somalia (Coleoptera). – Frustula entomologica N. S. 18 (31): 1–76; Firenze.
Ferrer, J. (2000a) Les espèces africaines appartenant au genre *Mesomorphus* Seidlitz 1893 (Coleoptera, Tenebrionidae, Opatrini). – Linzer biologische Beiträge 32/1: 463–527; Linz.
Ferrer, J. (2000b) Révision des espèces africaines et européennes appartenant au genre *Gonocephalum* Solier (Coleoptera, Tenebrionidae, Opatrini). Troisième partie. – Atti del Museo Civico di Storia naturale di Trieste 48: 69–153; Trieste.
Ferrer, J. (2005) Révision du genre *Opatroides* Brullé, 1832 (Coleoptera: Tenebrionidae: Opatrinae: Opatrini). – Annales Zoologici 55: 11–22; Warszawa. DOI: 10.3161/0003454053642248
Ferrer, J. (2010) Taxonomic Notes on the Genus *Gonocephalum* Solier, 1834, with Description of New Taxa (Coleoptera: Tenebrionidae). – Annales Zoologici 60: 231–238; Warszawa. DOI: 10.3161/000345410X516902
Ferrer, J. (2016) Contribución al conocimiento del género *Cheirodes* Géné [sic!] 1839, con descripción de nuevas especies de Europa y África (Coleoptera, Tenebrionidae). – Boletín de la Sociedad Entomológica Aragonesa (S.E.A) 59: 85–101; Zaragoza.
Ferrer, J., Castilla, A. M., Hawe, D., Abdulla, A. M. A. & Al-Hemaidi, A. A. M. (2014) Contributions to the knowledge of the genus *Scaurus* Fabricius, 1775 (Coleoptera,

Tenebrionidae) with description of new species from Qatar (*Scaurus qataricus* n. sp.) and from Cyprus (*Scaurus nielseni* n. sp.). – Boletín de la Sociedad Entomológica Aragonesa (S.E.A.) 55: 53–65; Zaragoza.

Ferrer, J., Martínez Fernández, J. C. & Castro Tovar, A. (2008) Aportación al conocimiento del género *Akis* Herbst, 1799 (Coleoptera, Tenebrionidae, Pimeliinae). – Boletín de la Sociedad Entomológico Aragonesa (S.E.A.) 43: 153–172; Zaragoza.

Fischer von Waldheim (1820) Entomographia Imperii Russici. – Auctoritate societas Caesareae Mosquensis naturae scrutatorum collecta et in lucem edita. Vol. 1., 26 pls., Augusti Semen; Moscow.

Fischer von Waldheim (1822) Entomographia Imperii Russici. Vol. 1. – Auctoritate societas Caesareae Mosquensis naturae scrutatorum collecta et in lucem edita. Vol. 1., viii + 210 pp, 25 pls.; Augusti Semen; Moscow.

Forskål, P. (1775) Descriptiones animalium avium, amphibiorum, piscium, insectorum, vermium, quae in itinere orientali observavit, post mortem edidit Carsten Niebuhr. 164 pp., 53 pls.; Kopenhavn.

Gebien, H. & Borchmann, F. (1927) Zur Erforschung des Persischen Golfes (Beitrag Nr. 14): Tenebrionidae (Col.) und Meloidae (Col.). – Supplementa Entomologica 16: 121–124; Berlin.

Gebien, H. (1906) Fam. Tenebrionidae. – In: Results of the Swedish Zoological Expedition to Egypt and the White Nile, under the direction of Leonard A. Jägerskiöld: 12–15; Library of the Royal University of Uppsala.

Gebien, H. (1937) Katalog der Tenebrioniden (Col. Heteromera). Teil I. – Pubblicazioni del Museo Entomologico "Pietro Rossi" 2: 505–883; Duino.

Gené, J. (1839) De quibusdam insectis sardiniae novis aut minus cognitis. Fasciculus II. – Memorie della Reale Academia della Scienze di Torino, serie secunda 1: 43–84, pls. 1-2; Torino.

Germar, E. F. (1836) Fauna Insectorum Europae. Fasciculus 18. Car. Aug. Kümmelii, 25 pp; Halae.

Girard, C. & Pierre, F. (1965) Sur quelques Tentyriini du Soudan (Republic of Sudan) (Col. Tenebrionidae). – Bulletin de la Société entomologique de France 1965: 134–137; Paris.

Gistel, J. (1848) Naturgeschichte des Thierreichs. Für höhere Schulen.Hoffmann'sche Verlags-Buchhandlung, xvi + 216; Stuttgart.

Goldberg, P. (1977) Late Quaternary stratigraphy of Gebel Maghara. – In: Bar-Yosef, O. & Phillips, J. L. (eds.): Prehistoric Investigations in Gebel Maghara, Northern Sinai. Qedem 7. – Monographs of the Institute of Archaeology. Hebrew University, 11–31; Jerusalem.

Gredler, V. (1878) Zur Käfer-Fauna Central-Afrikas. – Verhandlungen der kaiserlich-königlichen zoologisch-botanischen Gesellschaft in Wien 27 [1877]: 501–522; Wien.

Gridelli, E. (1928) Appunti su alcuni Tenebrionidae dell'Africa settentrionale. – Bollettino della Società Entomologica Italiana 6 (1–2): 15–18; Genova.

Gridelli, E. (1929a) Appunti su alcuni Tenebrionidae dell'Africa settentrionale. III. Revisione del genere *Doderoella*. – Bollettino della Società entomologica italiana 56 (1): 2–10; Genova.

Gridelli, E. (1929b) Appunti su alcuni Tenebrionidae dell'Africa settentrionale. IV. Revisione del genere *Tentyriina* Reitt. – Bollettino della Società entomologica italiana 56 (4): 49–63; Genova.

Gridelli, E. (1930) Risultati zoologici della missione inviata dalla R. Società Geographica Italiana per l'esplorazione dell'Oasi di Giarabub (1926-27). – Annali del Museo Civico di Storia Naturale di Genova 54: 1–485; Genova.

Gridelli, E. (1931) Spedizione del barone Raimondo Franchetti in Dancalia. Coleotteri – Tenebrionidae. – Annali del Museo Civico di Storia Naturale di Genova 55: 220–233; Genova.

Gridelli, E. (1933a) Spedizione scientifica all'oasi di Cufra. − Annali del Museo Civico di Storia Naturale di Genova 56: 155−258; Genova.
Gridelli, E. (1933b) Missione scientifica del prof. E. Zavattari nel Fezzan (1931). Coleotteri. − Bollettini della Società Entomologica Italiana 65: 70−90; Genova.
Gridelli, E. (1937) Coleotteri raccolti dal prof. G. Scortecci nel Fezzan (Missione R. Società Geografica 1934). − Atti della Società Italiana di Scienze Naturali e del Museo Civico di Storia Naturale di Milano 76: 17−54; Milano.
Gridelli, E. (1938) Coleotteri dell'Africa orientale italiana. 5. Contributo. Specie a me note apparenti ai generi *Mesostena* e *Mesostenopa* auct. (Tenebrionidae). − Bollettino della Società Entomologica Italiana 70: 103−121; Genova.
Gridelli, E. (1939a) Coleotteri dell'Africa orientale italiana. 8. contributo. Revisione delle specie riferite dagli autori ai generi *Helopinus*, *Drosochrus*, *Emyon* (Coleopt. Tenebrionidae). − Atti del Museo Civico di Storia Naturale di Trieste 14 (14): 187−206; Trieste.
Gridelli, E. (1939b) Coleotteri dell'Africa orientale italiana. 10. contributo. Revisione delle specie del genere *Leichenum* Blch. (Coleopt. Tenebrionidae). − Atti del Museo Civico di Storia Naturale di Trieste 14 (15): 207−242; Trieste.
Gridelli, E. (1939c) Coleoptera: Staphylinidae, Diversicornia, Heteromera, Lamellicornia, Chrysomelidae (partim). − In: Reale Accademia d'Italia, Centro Studi per l'Africa orientale (ed.): Missione Biologica nel paese dei Borana. Raccolte zoologiche, vol. II, parte 1: 311−615; Roma.
Gridelli, E. (1939d) Coleotteri del Fezzan e dei Tassili d'Aggér (Missione Scortecci 1936). − Atti della Società Italiana di Scienze Naturali e del Museo Civico di Storia Naturale in Milano 78: 385−456; Milano.
Gridelli, E. (1940) Coleotteri dell'Africa orientale italiana. 11. contributo. Materiali per lo studio della fauna eritrea raccolti nel 1901-03 dal dott. Alfredo Andreini. − Memorie della Società Entomologica Italiana 18 [1939]: 219−258; Genova.
Gridelli, E. (1948) Coleotteri dell'Africa Tropicale. XVII contributo. Ulteriori appunti per una monografia delle specie africane del genere *Gonocephalum* Sol. (Coleopt. Tenebr.). − Atti del Museo Civico di Storia Naturale di Trieste 17 (1): 1−56; Trieste.
Gridelli, E. (1950) Contribution à l'étude de l'Aïr (Mission L. Chopard et A. Villiers). Coléoptères Tenebrionidae. − Mémoires de l'Institut français d'Afrique noire 10: 153−180; Dakar.
Gridelli, E. (1952) Contribution à l'étude du peuplement de la Mauritanie. − Bulletin de l'Institut français d'Afrique noire 14: 60−96; Dakar.
Gridelli, E. (1953) Catalogo regionato delle specie di Coleotteri Tenebrionidi dell'Arabia. − Atti del Museo Civico di Storia Naturale di Trieste 19 (1): 1−70; Trieste.
Gridelli, E. (1954) Catalogo regionato delle specie della famiglia Tenebrionidae a me note dell'Afghanistan (Coleoptera). − Atti del Museo Civico di Storia Naturale di Trieste 19 (4): 167−292; Trieste.
Grimm, R. (1981) Die Fauna der Ägäis-Insel Santorin. Teil 2 (Insecta: Coleoptera). − Stuttgarter Beiträge zur Naturkunde, Serie A (Biologie) 348: 1−14; Stuttgart.
Grimm, R. (1983): Schwarzkäfer und andere Insekten. − In: Dittrich, P. (ed.): Biologie der Sahara, 2. erweiterte Auflage. Prof. Dr. O. Dittrich, pp. 107−141; Frankfurt a.M.
Grimm, R. (1985) Zur Kenntnis der Tenebrioniden aus Süditalien (Insecta: Coleoptera). − Stuttgarter Beiträge zur Naturkunde, Serie A (Biologie) 379: 1−32; Stuttgart.
Grimm, R. (1991) Tenebrioniden von der Insel Zypern (Insecta: Coleoptera). − Biocosme Mésogéen 8 (1): 15−49; Nice.
Grimm, R. (1995) Beitrag zur Kenntnis der Schwarzkäferfauna der Kanarischen Inseln (Coleoptera, Tenebrionidae). − Mitteilungen der Münchner Entomologischen Gesellschaft 85: 33−50; München.
Grimm, R. (2001) Zur Taxonomie von *Eurycaulus* Fairmaire, 1868, Subgenus *Ammotrypes* Fairmaire, 1879, stat. n. (*Scleronimon* Reitter, 1904, syn. n.) (Coleoptera, Tenebrionidae). − Nachrichtenblatt der bayerischen Entomologen 50 (3): 88−90; München.

Grimm, R. (2005) Taxonomic and faunistic notes on the genus *Eurycaulus*, with descriptions of two new species from the Arabian Peninsula (Coleoptera: Tenebrionidae). – Stuttgarter Beiträge zur Naturkunde, Serie A (Biologie) 672: 1–11; Stuttgart.

Grimm, R., Lillig, M. & Mifsud, D. (2020) An update for the Tenebrionidae (Coleoptera) of the Maltese Archipelago. – Bulletin of the entomological society of Malta 11: 47–56; Marsa. DOI: 10.17387/BULLENTSOCMALTA.2020.07

Guérin-Méneville, F. E. (1844) Iconographie du règne animal de G. Cuvier, ou représentation d'après nature de l'une des espèces les plus remarquables, et souvent non encore figurées, de chaque genre d'animaux. Avec un texte descriptif mis au courant de la science. Ouvrage pouvant servir d'atlas à tous les traités de zoologie. Insectes. [Texte.] J.-B. Baillière, 576 pp.; Paris.

Guérin-Méneville, F. E. (1862) Description d'une nouvelle espèce de Coléoptère du genre *Melancrus* et une rectification relative à une note publiée dans le Bulletin entomologique de 1859. – Annales de la Société entomologique de France (4. série) 1 [1861]: 375–376; Paris.

Haag-Rutenberg, G. (1870) Beiträge zur Familie der Tenebrioniden. – Coleopterologische Hefte 6: 84–93; München.

Haag-Rutenberg, G. (1875a) *Hidrosis*, eine neue Adelostomiden-Gattung. – Deutsche Entomologische Zeitschrift 19 (1): 120; Berlin.

Haag-Rutenberg, G. (1875b) Beiträge zur Familie der Tenebrioniden. – Coleopterologische Hefte 14: 67–92; München.

Haag-Rutenberg, G. (1876) Neue Pimeliden. – Biologische Monatsblätter 1: 75–76; Berlin.

Haag-Rutenberg, G. (1877) Revision der Gattung *Himatismus* Er. – Deutsche Entomologische Zeitschrift 21 (2): 273–284; Berlin.

Haag-Rutenberg, G. (1879) Fernere Nachträge zu den Heteromeren-Monographien der Moluriden, Eurychoriden und Adesmiiden. – Deutsche Entomologische Zeitschrift 23 (2): 289–296; Berlin.

Hafez, M. & Makki, A. M. M. (1959) Studies on desert insects in Egypt. III. On the bionomics of *Adesmia bicarinata* Klug. – Bulletin de la Société entomologique d'Égypte 52: 89–113; Cairo.

Hart, H. C. (1885) A Naturalist's Journey to Sinai, Petra, and South Palestine. – Palestine Exploration Fund, Quarterly Statement 1885: 231–286; London.

Hart, H. C. (1891) Some Account of the Fauna and Flora of Sinai, Petra, and Wâdy 'Arabah. 255 pp, 12 pls.; London.

Hassan, M. I., Bream, A. S., Younes, I. M., Elshewy, D. A., Khalifa, M. A. & Riad, S. A. (2017) Morphometric Comparison Between Western, Eastern Deserts and Sinai Populations of *Mesostena angustata* (Coleoptera: Tenebrionidae) in Egypt.– Egyptian Journal of Zoology 67: 51–66; Cairo. DOI: 10.12816/0037794

Herbst, J. F. W. (1797) Natursystem aller bekannten in- und ausländischen Insekten, als eine Fortsetzung der von Buffonschen Naturgeschichte. Der Käfer siebenter Theil. Geh. Commerzien-Raths Pauli 346 pp.; Berlin.

Herbst, J. F. W. (1799) Natursystem aller bekannten in- und ausländischen Insekten, als eine Fortsetzung der von Buffonschen Naturgeschichte. Der Käfer, achter Theil. Geh. Commerzien-Raths Pauli, 420 pp; Berlin.

Heyden, L. von (1899) Beitrag zur Coleopteren-Fauna der Halbinsel Sinaï. – Deutsche Entomologische Zeitschrift 43 (2): 240–256; Berlin.

Hinton, H. E. (1948) A Synopsis of the Genus *Tribolium* MacLeay, with some Remarks on the Evolution of its Species Groups (Coleoptera, Tenebrionidae). – Bulletin of entomological Research 39: 13–56; Cambridge.

Holdhaus, K. (1919) Koleopteren aus Mesopotamien. – Annalen des Naturhistorischen Museums Wien 33: 39–58, pl. 1; Wien.

Hoter, A., Rizk, S. & Naim, H. Y. (2019) Cellular and Molecular Adaptation of Arabian Camel to Heat Stress. Frontiers in genetics 10: 588; Lausanne. DOI: 10.3389/fgene.2019.00588

Innes Bey, W. (1909) Notice bibliographique de travaux entomologiques relatifs à l'Égypte. – Bulletin de la Société Entomologique d'Égypte 2: 72–86; Cairo.
Innes Bey, W. (1912) Une liste d'Insectes recueillis probablement par J. Lord en Égypte et déterminés par F. Walker. – Bulletin de la Société Entomologique d'Égypte 3 [1911]: 97–111; Cairo.
Iwan, D., Ferrer, J. & Ras, M. (2010) Catalogue of the World *Gonocephalum* Solier, 1834 (Coleoptera, Tenebrionidae, Opatrini). Part 1. List of the Species and Subspecies. – Annales Zoologici 60: 245–304; Warszawa.
Iwan, D., Löbl, I., Bouchard, P., Bousquet, Y., Kamiński, M., Merkl, O., Ando, K. & Schawaller, W. (2020) Family Tenebrionidae Latreille, 1802. – In: Iwan, D. & Löbl, I. (eds.): Catalogue of Palaearctic Coleoptera Volume 5, Tenebrionoidea. Revised and Updated Second Edition, I. Brill NV, 969 pp; Leiden. DOI: 10.3161/000345410X516920
Jacquelin du Val, P. N. C. (1861) Manuel entomologique. Genera des coléoptères d'Europe comprenant leur classification en familles naturelles, la description de tous les genres, des tableaux synoptiques destinés à faciliter l'étude, le catalogue de toutes les espèces, de nombreux dessins au trait de caractères et plus de treize cents types représentant un ou plusieurs insectes de chaque genre dessinés et peints d'après nature avec le plus grand soin par M. Jules Migneaux. Tome troisième. A. Deyrolle, pp. 273–352; Paris.
Kamiński, M. J., Lumen, R., Kanda, K., Iwan, D., Johnston, M. A., Kergoat, G., Bouchard, P., Bai, X.-L., Li, X.-M., Ren, G.-D. & Smith, A. D. (2020) Reevaluation of Blapimorpha and Opatrinae: addressing a major phylogeny-classification gap in darkling beetles (Coleoptera: Tenebrionidae: Blaptinae). – Systematic Entomology 46, 140–156; London. DOI: 10.1111/syen.12453
Kaszab, Z. (1967) Die Arten der Gattung *Philhammus* Fairm. (Coleoptera, Tenebrionidae). – Annales historico-naturales Musei nationalis hungarici, pars zoologica 59: 291–296; Budapest.
Kaszab, Z. (1969) Beiträge zur Kenntnis der Fauna Afghanistans. Tenebrionidae, Col. (Fortsetzung). – Acta Musei Moraviae 54 (Supplementum): 5–182; Brno.
Kaszab, Z. (1972): Missione 1965 del Prof. Guiseppe Scortecci nelle Yemen (Arabian meridienale). Coleoptera Tenebrionidae. – Atti della Società Naturali et Museo Civico di Storia Naturale di Milano 113 (4): 366–384, Milano.
Kaszab, Z. (1979) Insects of Saudi Arabia. Coleoptera: Fam. Tenebrionidae. – Fauna of Saudi Arabia 1: 257–288; Basel.
Kaszab, Z. (1981) Insects of Saudi Arabia. Coleoptera: Fam. Tenebrionidae (Part 2). – Fauna of Saudi Arabia 3: 276–401; Basel.
Kaszab, Z. (1982) Insects of Saudi Arabia. Coleoptera: Fam. Tenebrionidae (Part 2). – Fauna of Saudi Arabia 4: 124–243; Basel.
Kaszab, Z. (1959) Wissenschaftliche Ergebnisse der zoologischen Expedition des National-Museums in Prag nach der Türkei. 24. Coleoptera Tenebrionidae. – Acta Entomologica Musei Nationalis Pragae 33: 69–82; Praha.
Katbeh-Bader, A. (1996) Contribution to our knowledge of the Tenebrionidae (Coleoptera) of Jordan. – Zoology of the Middle East 13: 99–106; Heidelberg. DOI: 10.1080/09397140.1996.10637711
Kirsch, T. F. W. (1869) Beitrag zur Kenntnis der Gattung *Omophlus* Sol. – Berliner entomologische Zeitschrift 13: 97–128; Berlin.
Klug, J. C. F. (1830) Symbolae physicae seu icones et descriptiones insectorum: quae ex itinere per Africam Borealem et Asiam Occidentalem Friderici Guilelmi Hemprich et Christiani Godofredi Ehrenberg medicinae et chirurgiae doctorum studio novae aut illustratae redierunt. Decas secunda. [29] pp, pls. XI-XX. Officina Academica; Berlin.
Kneucker, A. (1903) Zoologische Ausbeute einer botanischen Studienreise durch die Sinai-Halbinsel im März und April 1902. – Verhandlungen der kaiserlich-königlichen zoologisch-botanischen Gesellschaft in Wien 53: 575–587; Wien.

Koch, C. (1934) Beitrag zur Kenntnis der Tenebrioniden vom Gebel Elba und Mersa Halaib. – Bulletin de la Société Royale Entomologique d'Égypte 18 (1–2): 92–103; Cairo.

Koch, C. (1935a) Wissenschaftliche Ergebnisse der entomologischen Expedition Seiner Durchlaucht des Fuersten A. della Torre e Tasso nach Aegypten und auf die Halbinsel Sinai. – Bulletin de la Société Royale Entomologique d'Égypte 19: 2–111; Cairo.

Koch, C. (1935b) Wissenschaftliche Ergebnisse der entomologischen Expedition Seiner Durchlaucht des Fuersten Alexander C. della Torre e Tasso in Palaestina. I. Tenebrionidae (Coleoptera). – Bulletin de la Societé Royale Entomologique d'Égypte 19: 262–288; Cairo.

Koch, C. (1937) Wissenschaftliche Ergebnisse über die während der Expedition Seiner Durchlaucht des Fürsten Alessandro C. della Torre e Tasso in Lybien aufgefundenen Tenebrioniden. – Pubblicazioni del Museo Entomologico "Pietro Rossi" 2: 285–500; Duino.

Koch, C. (1940a) Über einige neue und interessante, von Johannes Roth 1857 im Wadi Arabah gesammelte Tenebrioniden (Col.). – Mitteilungen der Münchner Entomologischen Gesellschaft 30: 84–107; München.

Koch, C. (1940b) Phylogenetische, biogeographische und systematische Studien über ungeflügelte Tenebrioniden (Col. Tenebr.). – Mitteilungen der Münchner Entomologischen Gesellschaft 30: 254–337; München.

Koch, C. (1940c) Der saharo-sindische Verbreitungs-Typus bei der ungefluegelten Tenebrioniden-Gattung *Mesostena* unter Berücksichtigung ähnlicher biogeographischer Verhältnisse der *Tentyria*-Stamm-Gruppe *Tentyrina* (Col. Tenebr.). – Rivista Biologica Coloniale 3 (1–2): 1–137; Roma.

Koch, C. (1941a) Contributo alla conoscenza del genere *Mitotagenia* Rtt. (Col., Tenebr.). – Memorie della Società Entomologica Italiana 20: 37–42; Genova.

Koch, C. (1941b) Phylogenetische, biogeographische und systematische Studien über ungeflügelte Tenebrioniden (Col. Tenebr.). – Mitteilungen der Münchner Entomologischen Gesellschaft 31: 252–314; München.

Koch, C. (1944) Tenebrionidae I (I. Beitrag zur Kenntnis der iberischen Fauna.). – Mitteilungen der Münchner Entomologischen Gesellschaft 34: 216–254; München.

Koch, C. (1949) Die beiden Rassenkreise der *Adesmia biskrensis* und *montana*. – EOS: Revista Española de Entomología 25: 115–144; Madrid.

Koch, C. (1950) The Tenebrionidae of Southern Africa. I. First Account of the Tenebrionidae collected on the University of California-Transvaal Museum Expedition, 1948. – Annals of the Transvaal Museum 21 (3): 273–376, pls. V-XXI.; Pretoria.

Koch, C. (1951) Beitrag zur Kenntnis der Gattung *Adesmia* (Col. Tenebrionid.). – Entomologische Blätter 45–46 [1949–1950]: 121–130; Krefeld.

Koch, C. (1965) Missione 1962 del Prof. Giusseppe Scortecci nell'Arabia meridionale. Coleoptera Tenebrionidae. Includendo materiale di viaggi in Arabia del Sig. G. Popov (1962) e del Dr. G. Bernadelli (1962-63). – Atti della Società Italiana della Scienze Naturali e del Museo Civico di Storia Naturale di Milano 104: 99–154, pls. V-VII; Milano.

Kocher, L. (1950) Notes sur les *Erodius* marocains (Contribution à l'étude systématique du genre). – Bulletin de la Société des Sciences naturelles du Maroc 30: 195–230; Paris.

Kocher, L. (1955) Note complémentaire sur les *Erodius* marocains. – Bulletin de la Société des Sciences naturelles et physiques du Maroc 35: 243–249; Rabat.

Kocher, L. (1958) Catalogue commenté des Coléoptères du Maroc 6, Ténébrionides. – Travaux de l'Institut Scientifique Chérifien, Série Zoologie 12: 1–185; Rabat.

Kocher, L. (1964) Catalogue commenté des Coléoptères du Maroc 10, Addenda et Corrigenda, Tables. – Travaux de l'Institut Scientifique Chérifien, Série Zoologie 30: 1–200; Rabat.

Kraatz, G. (1865) Revision der Tenebrioniden aus der alten Welt aus Lacordaire's Gruppen der Erodiides, Tentyriides, Akisides, Piméliides und der europäischen *Zophosis*-Arten. 393 pp.; Berlin

Krasnov, B. & Ayal, Y. (1995) Seasonal changes in darkling beetle communities (Coleoptera: Tenebrionidae) in the Ramon erosion cirque, Negev Highlands, Israel. – Journal of Arid Environments 31: 335–347; London. DOI: 10.1016/S0140-1963(05)80037-0

Kühnelt, W. (1965) Catalogus Faunae Graeciae. Pars 1. Tenebrionidae. – To Vouno 1965: 1–60; Athens.

Kuss, J. & Bachmann, M. (1996) Cretaceous paleogeography of the Sinai Peninsula and neighbouring areas. – Compte rendu de l'Académie des Sciences de Paris, t. 322, ser. II a: 915–933; Paris.

Küster, H. C. (1848) Die Käfer Europas. Nach der Natur beschrieben. Heft 13: 2 + 100 pp, 3 pls.; Nürnberg.

Küster, H. C. (1849) Die Käfer Europas. Nach der Natur beschrieben. Heft 16: 4 + 100 pp, 3 pls.; Nürnberg.

Kwieton, E. (1977) Révision phylogénétique du groupe de *Pimelia granulata* (Coleoptera, Tenebrionidae). – Annotationes zoologicae et botanicae 121: 1–29; Bratislava.

Kwieton, E. (1978) Essai synoptique sur le genre *Thriptera* Solier (Col. Tenebrionidae). – Annotationes Zoologicae et Botanicae 123: 1–15; Bratislava.

Kwieton, E. (1980) Synopsis des espèces du genre *Sepidium* F. d'Algérie et de Tunisie (Col., Tenebrionidae). – Annotationes zoologicae et botanicae 138: 1–19; Bratislava.

Kwieton, E. (1981) Coléoptères Tenebrionidae constatés par l'expédition de TIS au Pakistan. – Annotationes Zoologicae et Botanicae 143: 1–18; Bratislava.

Kwieton, E. (1982) Contributions ultérieures à la connaissance de genre *Pimelia* F. (Col., Tenebrionidae). – Annotationes Zoologicae et Botanicae 145: 1–38; Bratislava.

Lacordaire, J. T. (1859) Histoire naturelle des insectes. Genera des coléoptères ou exposé méthodique et critique de tous les genres proposés jusqu'ici dans cet ordre d'insectes. Tome V, première partie. Librairie Encyclopédique de Roret: 1–400; Paris.

Lapeva-Gjonova, A. (2013) Ant-Associated Beetle Fauna in Bulgaria: A Review and New Data – Psyche 2013, Article ID 242037, 14 pages; New York. DOI: 10.1155/2013/242037

Laporte de Castelnau, F. L. N. C. (1840) Histoire naturelle des insectes coléoptères; avec une introduction renfermant l'anatomie et la physiologie des animaux articulés, par M. Brullé. Tome deuxième. P. Duménil, 563 pp., 38 pls.; Paris.

Larsen, T. B. (1990) The Butterflies of Egypt. Apollo Books, 112 pp.; Svendborg.

Latreille P. A. (1802) Histoire naturelle, générale et particulière, des crustacés et des insectes. Ouvrage faisant suite à l'histoire naturelle générale et particulière, composée par Leclerc de Buffon, et rédigée par C. S. Sonnini, membre de plusieurs sociétés savantes. Familles naturelles des genres. Tome troisième. F. Dufart, i-xii, 13–46; Paris.

Latreille, P. A. (1809) Genera crustaceorum et insectorum secundum ordinem naturalem in familias disposita, iconibus exemplisque plurimis explicita. Tomus quartus et ultimus. Amand Koenig. 399 pp.: Parisiis et Argentorati.

Latreille, P. A. (1825) Familles naturelles du règne animal, exposées succinctement et dans un ordre analytique, avec l'indication de leurs genres. J.-B. Baillière, 570 pp.; Paris.

Latreille, P. A. (1829) – In: Cuvier, G.: Le règne animal distribué d'après son organisation, pour servir de base à l'histoire naturelle des animaux et d'introduction à l'anatomie comparée. Nouvelle édition, revue et augmentée. Tome V. Suite et fin des insectes. Déterville, xxii+ 556 pp.; Paris.

Leach, W. E. (1815) Entomology. Pp. 57–172. In: Brewster, D. (ed.) The Edinburgh encyclopedia. Volume 9. William Blackwood; Edinburgh.

Leo, P. (2020) Osservazioni sui taxa del gruppo di *Pimelia* (*Pimelia*) *angulata* Fabricius, 1775, con descrizione di una nuova specie del Marocco sud-orientale (Coleoptera, Tenebrionidae, Pimeliini). – Giornale italiana di Entomologia 15 (64): 809–830; Cremona.

Lillig, M. (1994) The Subgenus *Eodirosis* Kwieton of the Genus *Erodius* Fabricius, with Two New Species and a Key to Species (Coleoptera: Tenebrionidae, Pimeliinae, Erodiini). – Israel Journal of Entomology 28: 151–158; Jerusalem.

Bibliography

Lillig, M. (1995) Die Gattung *Scaurus* Fabricius, 1775 im Sudan. – Mitteilungen der Münchner Entomologischen Gesellschaft 85: 51–55; München.

Lillig, M. (1997) A new species of the Genus *Erodius* from Israel and Egypt (Coleoptera: Tenebrionidae: Pimeliinae). – Israel Journal of Entomology 31: 55–58; Jerusalem.

Lillig, M. (2001) Revision des Subgenus *Symphoxycara* Koch, 1943 der Gattung *Oxycara* Solier, 1835 (Coleoptera, Tenebrionidae, Tentyriini). – Coleoptera 5: 363–389; Schwanfeld.

Lillig, M. (2009) New species of the genus *Oxycara* Solier, 1835 from Oman (Coleoptera, Tenebrionidae). – Caucasian Entomological Bulletin 5 (2): 227–230; Rostov-on-Don. DOI: 10.23885/1814-3326-2009-5-2-227-230

Lillig, M. (2021) *Clitobius grimmi* sp. n. from Egypt with notes on *C. oblongiusculus* (Fairmaire, 1875) (Coleoptera: Tenebrionidae: Blaptinae). – Zootaxa 5082 (5): 476–484; Auckland. DOI: 10.11646/zootaxa.5082.5.5

Lillig, M., Borg Barthet, H. & Mifsud, D. (2012) An identification and informative guide to the Tenebrionidae of Malta (Coleoptera). – Bulletin of the entomological society of Malta 5: 121–160; Marsa.

Lillig, M. & Bremer, H. J. (2002): Tenebrionidae der nördlichen Provinzen der Republik Sudan (Coleoptera: Tenebrionidae). – Coleoptera 6: 35–90; Schwanfeld.

Lillig, M. & Pavlíček, T. (2002) *Astorthocnemis becvarorum*, a new genus and a new species from the Middle East (Coleoptera: Tenebrionidae, Pimeliinae, Platyopini). – Mitteilungen des Internationalen Entomologischen Vereins 28 (3/4): 97–104; Frankfurt a.M.

Lillig, M. & Pavlíček, T. (2003) The Darkling Beetles of the Sinai Pensinsula (Coleoptera: Tenebrionidae excl. Lagriinae et Alleculinae). – Zoology in the Middle East, Supplementum. Kasparek-Verlag, 87 pp., 2 pls; Heidelberg.

Lindberg Har. (1950) Beitrag zur Kenntnis der Käferfauna der Kanarischen Inseln. – Societas Scientiarum Fennica, Commentationes Biologicae 10 (18): 1–20, Helsinki.

Linnaeus, C. (1764) Museum S:ae R:ae Ludovicae Ulricae Reginae. Svecorum, Gothorum, Vandalorumque & c. &c. &c. In quo Animalis rariora, exotica, imprimis Insecta & Conchilia describuntur & determinantur. Prodromus instar editum. L. Salvii, vi + 720 + (2); Stockholm.

Löbl, I. & Merkl, O. (2003) On the type species of several Tenebrionid genera and subgenera (Coleoptera, Tenebrionidae). – Acta Zoologica Academiae Scientiarum Hungaricae 49 (3): 243–253; Budapest.

Lucas, P. H. (1844) Notes sur quelques nouvelles espèces d'insectes qui habitent les possessions françaises du nord de l'Afrique. – Revue zoologique 1844: 262–267; Paris.

Lucas, P. H. (1846) Pp. 1-360. – In: Exploration scientifique de l'Algérie pendant les années 1840, 1841, 1842 publiée par ordre du gouvernement et avec le concours d'une Commission Académique. Sciences physiques Zoologie. Vol. II. Histoire naturelle des animaux articulés. Cinquième classe. Insectes. Premier ordre. Les coléoptères. Imprimerie Nationale, 590 pp, 47 pls.; Paris.

Lucas, P. H. (1855) Note sur quelques nouvelles espèces de Coléoptères qui habitent les possessions françaises du nord de l'Afrique. – Revue et magazine de zoologie pure et appliquée (2. série) 7: 290–292; Paris.

Lucas, P. H. (1856) [New taxa]. Bulletin de la Société entomologique de France 1856: xlv-xlvi; Paris.

Lucas, P. H. (1858) La description de plusieurs Mélanosomes nouveaux qui habitent le sud des possessions françaises, dans le nord de l'Afrique. – Bulletin entomologique 1858: ccxx–ccxxiii; Paris.

MacLeay, W. S. (1825) *Annulosa* Javanica, or attempt to illustrate the natural affinities and analogies of the insects collected in Java by Thomas Horsfield, M.D.F.L. & G.S. and deposited by him in the honourable East-India Company. – Kingsbury, Parbury, and Allen, xii + 9–50 pp, 1 pl; London.

Makhan, D., Ezzatpanah, S., Ghahari, H., Naderian, H., Hadian, A. & Hawkeswood, T. J. (2011) *Mesostena angustata* (Fabricius, 1775) (Coleoptera: Tenebrionidae) from Semnan, Iran, feeding on ants of *Messor intermedius* Santschi, 1927 (Hymenoptera: Formicidae). – Calodema 171: 1–6; North Richmond.

Marseul, S. A. (1867) Descriptions d'espèces nouvelles. – L'Abeille 4: xxxiii–xl; Paris.

Marseul, S. A. (1876) Mélanges. Nouvelles et faits divers 2 (9). – L'Abeille 16: 35–36; Paris.

Matthews, E. G. (2004) New synonymy and new names in Australian Tenebrionidae (Coleoptera). – Transactions of the Royal Society of South Australia 128: 261; Adelaide.

Medvedev, G. S. & Nepesova, M. G. (1985) Opredelitel zhukov-chernotelok Tukmenistana. Ylym, 180pp.; Ashkhabad. [in Russian]

Ménétries, E. (1849) Catalogue des Insectes recueillis par feu M. Lehmann, avec les descriptions des nouvelles espèces. (Seconde et dernière partie). Coléoptères Hétéromères. – Mémoires de l'Académie Impériale des Sciences de St.-Petersbourgh, Sixième série, 6: 217–328, pls III-V; St. Petersburg.

Miedel, J. (1880) Observations sur les *Opatrum*. – Deutsche Entomologische Zeitschrift 24 (1): 136–140; Berlin.

Mifsud, D. & Scupola, A. (1998) The Tenebrionidae (Coleoptera) of the Maltese Islands (Central Mediterranean). – Annali del Museo Civico di Storia Naturale "G. Doria" 92: 191–229; Genova.

Miller, L. (1858) Zwei neue Erodiinen-Genera. – Wiener Entomologische Monatsschrift 2: 115–124; Wien.

Miller, L. (1861) Neue Käfer aus Kindermann's Vorräthen. – Wiener Entomologische Monatsschrift 5 (6): 169–182 & 5 (7): 201–209, pls. 4, 5; Wien.

Mulsant, E. & Rey, C. (1853) Essai d'une division des derniers Mélasomes. – Opuscules Entomologiques 4: 1–235, 4 pls.; Lyon.

Mulsant, E. & Rey, C. (1854) Essai d'une division des derniers Mélasomes (suite). Pandarites. Mémoires de l'Académie Impériale des Sciences, Belles-lettres et Arts de Lyon. Classe des Sciences 4: 153–332; Lyon.

Mulsant, E. & Rey, C. (1859) Essai d'une division des derniers Mélasomes, famille des Parvilabres. Quatrième tribu Opatrites. – Opuscules Entomologiques 10: 1–160; Paris.

Nabozhenko, M. V., Labrique, H., Korkuta, A. & Keskin, B. (2020) Contribution to the knowledge of the genus *Scaurus* Fabricius, 1775 (Coleoptera: Tenebrionidae: Scaurini) from Turkey. – Caucasian Entomological Bulletin 16 (1): 141–151; Rostov-on-Don. DOI: 10.23885/181433262020161-141151

Nabozhenko, M. V. & Lillig, M. (2013) A new subgenus and species of the genus *Hedyphanes* Fischer von Waldheim, 1820 (Coleoptera: Tenebrionidae: Helopini) from Israel and Egypt. – Zootaxa 3641 (2): 188–192; Auckland. DOI: 10.11646/zootaxa.3641.2.6

Norfolk, O., Abdel-Dayem, M. & Gilbert, F. (2012) Rainwater harvesting and arthropod biodiversity within an arid agro-ecosystem. – Agriculture, Ecosystems & Environment 162: 8–14; Amsterdam. DOI: 10.1016/j.agee.2012.08.007

Normand, H. (1936) Contribution au Catalogue de Coléoptères de la Tunisie. 10e Fascicule. – Bulletin de la Société d'Histoire Naturelle d'Afrique du Nord 27: 355–383; Alger.

Olivier, G.-A. (1791) Encyclopédie méthodique, ou par ordre de matières. Histoire Naturelle. Insectes. Tome sixième. Panckoucke, 704 pp; Paris.

Olivier, G.-A. (1795) Entomologie, ou histoire naturelle des insectes, avec leurs caractères génériques et spécifiques, leur description, leur synonymie, et leur figure enluminée. Coléoptères. Tome troisième. Imp. de Lanneau; Paris.

Olivier, G.-A. (1811) Encyclopédie méthodique, ou par ordre de matières. Histoire Naturelle. Insectes. Tome huitième. H. Agasse, 722 pp.; Paris.

Oromí, P. (1982) Los Tenebriónidos de las Islas Canarias. Instituto de Estudios Canarios 50 Aniversario (1932-1982). Tomo 1: 267–292; Santa Cruz de Tenerife.

Panzer, G. W. F. (1796) Entomologia Germanica initia oder Deutschlands Insecten. Heft 37, 24 pp. + 24 pls, Felsecker; Nürnberg.
Pavlíček, T., Csuzdi, Cs. & Nevo, E. (1997) The first recorded earthworms from Negev and Sinai deserts. Israel Journal of Zoology 43 (1): 1–3; Jerusalem. DOI: 10.11646/zootaxa.3641.2.6
Penrith, M.-L. (1982a) Revision of the Zophosini (Coleoptera: Tenebrionidae). Part 5. A derived subgenus from northern Africa. – Cimbebasia (A), 6 (5): 165–226; Windhoek.
Penrith, M.-L. (1982b) Revision of the Zophosini (Coleoptera: Tenebrionidase). Part 6. The subgenus *Hologenosis* Deyrolle. – Cimbebasia (A) 6 (6): 228–289; Windhoek.
Penrith, M.-L. (1983) Revision of the Zophosini (Coleoptera: Tenebrionidae). Part 8. The palaearctic species group of the subgenus *Oculosis* Penrith, the subgenus *Cheirosis* Deyrolle, and a monotypical subgenus from Socotra. – Cimbebasia (A), 6 (9): 369–384; Windhoek.
Penrith, M.-L. (1984) Revision of the Zophosini (Coleoptera: Tenebrionidae). Part 9. The remaining species of the subgenus *Septentriophosis* Penrith. – Cimbebasia (A), 6 (9): 365–416; Windhoek.
Penrith, M.-L. (1986) Revision of the Zophosini (Coleoptera: Tenebrionidae). Part 10. Key to the subgenera, supplement, evolution and biogeography of the tribe, and catalogue. – Cimbebasia (A), 6 (10): 417–502; Windhoek.
Peyerimhoff, P. de (1907) Liste des Coléoptères du Sinaï. – L'Abeille 31: 1–48; Paris.
Peyerimhoff, P. de (1931) Mission scientifique du Hoggar. Coléoptères. – Mémoires de la Société d'Histoire Naturelle d'Afrique du Nord 2: 5–173; Alger.
Peyerimhoff, P. de (1933) Description de quatre Tenebrionidae (Col.) du Sahara central. – Bulletin de la Société d'Histoire naturelle de l'Afrique du Nord 24: 344–356; Alger.
Peyerimhoff, P. de (1934) Mission transsaharienne de S. A. R. le Prince Sixte de Bourbon-Parme (1932). Insectes Coléoptères. – Revue française d'Entomologie 1: 104–108; Paris.
Peyerimhoff, P. de (1935) Coléoptères marocains récoltés par M. Ch. Rungs. – Bulletin de la Société des Sciences naturelles du Maroc 15: 19–25; Rabat.
Peyerimhoff, P. de (1936) Coléoptères (avec la description d'un genre et de deux espèces nouvelles). – In: Dalloni, M. (ed.): Mission au Tibesti (1930-31). – Mémoires de l'Académie des Sciences de l'Institut de France 62: 74–81; Paris.
Peyerimhoff, P. de (1944) Coléoptères du Sahara marocain et du Sahara occidental. – Bulletin de la Société des Sciences naturelles du Maroc 24: 90–110; Rabat.
Peyerimhoff, P. de (1948a) Insectes Coléoptères. – Mission scientifique au Fezzân (1944-1945). 5 (Zoologie): 6–84; Alger.
Peyerimhoff, P. de (1948b) Étude sur la systématique des Coléoptères du nord africain. II. Les *Scaurus* (Tenebrionidae). – Revue française d'Entomologie 14 (3): 157–193; Paris.
Pic, M. (1921) *Fourtaus* N. Gen. Pic. – In: Alfieri, A. Contributions à la faune entomologique du Nord-Est du Sinaï. – Bulletin de la Société entomologique d'Égypte 6 [1919-1020]: 40–53; Cairo.
Pic, M. (1923a) Notes diverses, descriptions et diagnoses (Suite). – L'Échange, Revue Linnéenne 39 (412): 5–7; Moulins.
Pic, M. (1923b): Notes diverses. Descriptions et diagnoses (Suite). – L'Échange, Revue Linnéenne 39 (412): 9–11; Moulins.
Pic, M. (1936) Nouveaux coléoptères exotiques. – Bulletin de la Société zoologique de France 61: 298–300; Paris.
Pic, M. (1939) Coléoptères nouveaux d'Égypte et du Sinaï. – Bulletin de la Société Fouad 1er d'Entomologie 23: 143–149; Cairo.
Pierre, F. (1958) Écologie et peuplement entomologique des sables vifs du Sahara nord-occidental. – Centre National de la Recherche Scientifique. 332 pp., Paris.
Pierre, F. (1961a) Écologie et distribution géographique des *Storthocnemis* Karsch. Affinités. (Tenebrionidae). 11. Internationaler Kongress für Entomologie in Wien 1960 1: 555–558; Wien.

Pierre, F. (1961b) Les Ténébrionides du Tibesti et du Borkou (Missions P. de Miré et P. Quézel). – Bulletin de l'Institut Fondamental d'Afrique Noire 23, A 4: 1030–1053; Dakar.
Pierre, F. (1963) Sur quelques Ténébrionides recueillis au Fezzan par le professeur A. Pasa. – Memorie del Museo Civico di Storia naturale di Verona 11: 205–209; Verona.
Pierre, F. (1972) Les *Caedius* du continent africain. Écologie, biogéographie et taxonomie (Tenebrionidae). – Annales de la Société entomolgique de France (N. S.) 8 (4): 951–981; Paris.
Pierre, F. (1979) Les Tenebrionidae du Djebel Marra (Soudan) et notes sur quelques particularités de leur morphologie. – Bulletin de la Société entomologique de France 84: 4–10; Paris.
Pomel, A. (1871) Le Sahara (Géologie, Géographie et Biologie). – Bulletin de la Société Algérienne de Climatologie. Sciences Physiques et Naturelles 2 (8): 133–265; Alger.
Pope, R. D. & Watt, J. C. (1986) *Tribolium castaneum* (Herbst, 1797) (Insecta, Coleoptera): proposed conservation by the suppression of *Tribolium navale* (Fabricius, 1775). Z.N.(S.) 2575. – The Bulletin of zoological nomenclature 43: 363–365; London.
Pospischil, R. (1996): Der glänzendschwarze Getreideschimmelkäfer. – Der Praktische Schädlingsbekämpfer 48 (12): 14–15; Lehrte.
Purchart, L. (2007) *Adelostoma* (*Zarudnyomus*) *borowieci* n. sp. from India (Coleoptera: Tenebrionidae: Adelostomini). – Genus 18 (2): 239–245; Wrocław.
Purchart, L. & Kamiński, M. J. (2017) A taxonomic review of the genus *Clitobius* with description of a new species from Oman (Coleoptera: Tenebrionidae). – Acta entomologica Musei nationalis Pragae 57 (supplementum); 139–163; Prague. DOI: 10.1515/aemnp-2017-0115
Raffray, A. (1873) *Anemia? submetallica*, R. – Revue et Magazine de Zoologie Pure et Appliquée (3) 1: 377–378, pl. 8; Paris.
Redtenbacher, L. (1868) Coleopteren. – In: Wüllerstorf-Urbair, B. (ed.) Reise der österreichischen Fregatte Novara um die Erde. – Zoologischer Theil. 2, Abtheilung 1. A, 249 pp., 5 pls; Wien.
Reichardt, A. N. (1936) Révision des Opatrines (Coleoptera Tenebrionidae) de la région paléarctique. Tableaux Analytiques de la Faune de l'URSS. – Zoologie 19: 1–224; Moscow & Leningrad. [in Russian]
Reiche, L. J. (1847) Entomologie. Pp. 257–532. – In: Ferret, P. & Galain, J. G. (eds.): Voyage en Abyssinie dans les provinces du Tigré, du Samen et de l'Amhara 3 [1847-1848]. Paulin, 536 pp.; Paris.
Reiche, L. J. (1860) Description d'une nouvelle espèce de Coléoptère du genre *Microtelus*. – Annales de la Société entomologique de France 8 (3. série): 334; Paris.
Reiche, L. J. (1861) Sur quelques espèces du Nord de l'Afrique. – Annales de la Société entomologique de France 1 (4. série): 87–92; Paris.
Reiche, L. J. & Saulcy, F. de (1857) Espèces nouvelles ou peu connues de Coléoptères, recueillies par M. F. de Saulcy, membre de l'Institut, dans son Voyage en Orient. – Annales de la Société entomologique de France 5 (3. série): 169–276, pl. 5; Paris.
Reitter, E. (1884) Ueber die syrischen Arten der Gattung *Anemia* Lap. – Deutsche Entomologische Zeitschrift 28: 259–260; Berlin.
Reitter, E. (1893) Bestimmungs-Tabelle der unechten Pimeliden aus der palaearctischen Fauna. – Verhandlungen des Naturforschenden Vereins in Brünn 31 [1892]: 201–250; Brünn.
Reitter, E. (1894) Fünfter Beitrag zur Coleopteren-Fauna von Europa und den angrenzenden Ländern. – Wiener Entomologische Zeitung 13: 299–306; Wien.
Reitter, E. (1897) Ueber die Arten der Coleopteren-Gattung: *Anemia* Cast., aus der palaearctischen Fauna. – Fest-Schrift zur Feier des fünfzigjährigen Bestehens des Vereins für schlesische Insectenkunde in Breslau 1847–1897: 40–43; Breslau.

Reitter, E. (1900) Bestimmungs-Tabelle der Tenebrioniden-Abtheilungen: Tentyrini und Adelostomini aus Europa und den angrenzenden Ländern. – Verhandlungen des naturforschenden Vereins in Brünn 39: 82–197, Brünn.

Reitter, E. (1904) Bestimmungs-Tabelle der Tenebrioniden-Unterfamilien Lachnogyini, Akidini, Pedinini, Opatrini und Trachyscelini aus Europa und den angrenzenden Ländern. – Verhandlungen des Naturforschenden Vereins in Brünn 42 [1903]: 25–189; Brünn.

Reitter, E. (1907) Ein neuer *Microtelus* (Sol.) aus Aegypten (Coleoptera, Tenebrionidae). – Wiener Entomologische Zeitung 26: 115; Wien.

Reitter, E. (1910) *Psilachnopus Aharonii* n. sp. – Wiener Entomologische Zeitung 29: 197; Wien.

Reitter, E. (1914a) Bestimmungs-Tabelle für die Unterfamilie Erodini der Tenebrionidae, aus Europa und den angrenzenden Ländern. – Deutsche Entomologische Zeitschrift 1914: 43–85; Berlin.

Reitter, E. (1914b) Bestimmungs-Tabelle der Tenebrioniden-Abteilung der Sepidiini. – Deutsche Entomologische Zeitschrift 1914: 381–392, Berlin.

Reitter, E. (1914c) Bestimmungs-Tabelle der Tenebrioniden-Abteilung der Scaurini. – Deutsche Entomologische Zeitschrift 1914: 369–380; Berlin.

Reitter, E. (1916a) Bestimmungstabelle der Arten der Gattung *Adesmia* Fisch. aus der paläarktischen Fauna (Col. Tenebrionidae– Wiener Entomologische Zeitung 35 (1–2): 1–31; Wien.

Reitter, E. (1916b) Bestimmungstabelle der Tenebrioniden-Gruppe der Phaleriini, aus der palaearktischen Fauna. – Entomologische Blätter 1916 (1–3): 3–10; Krefeld.

Reitter, E. (1916c) Bestimmungstabelle der Tenebrioniden-Unterfamilie Zophosini aus der paläarktischen Fauna. – Wiener Entomologische Zeitung 35 (3–4): 81–99; Wien.

Reitter, E. (1917) Bestimmungs-Schlüssel für die Unterfamilien und Tribus der paläarktischen Tenebrionidae. – Wiener Entomologische Zeitung 36 (3–5): 51–66; Wien.

Renan, I., Freidberg, A., Groner, E. & Bar Kutiel, P. (2020) Impact of spatial patterns on arthropod assemblages following natural dune stabilization under extreme arid conditions. – Global Journal of Ecology 5 (1): 79–87; Los Angeles. DOI: 10.17352/gje.000024

Reymond, A. (1952) Insectes de divers ordres récoltés au Sahara central au cours d'une mission du Centre National de la Recherche scientifique en 1947-1948. – Bulletin de la Société des Sciences naturelles du Maroc 32: 77–89; Rabat.

Rougon, D. & Ardoin, P. (1976) Contribution à l'étude de la faune entomologique de la République du Niger. III. Premier inventaire des Tenebrionidae (Coleoptera). – Bulletin de l'Institut fondamental d'Afrique noire, 38 (2), série A: 303–341; Dakar.

Saunders, D. (2021) Insect photoperiodism: Büning's hypothesis, the history and development of an idea. –European Journal of Entomology 118: 1–13; České Budějovice. DOI: 10.14411/eje.2021.001

Schatzmayr, A. (1938) Coleotteri raccolti dal dott. Edgardo Moltoni durante le sue missioni ornitologiche in Libia. – Atti della Società Italiana di Scienze Naturali e del Museo Civico di Storia Naturale di Milano 77: 159–190; Milano.

Schatzmayr, A. & Koch, C. (1934) Wissenschaftliche Ergebnisse der entomologischen Expedition Seiner Durchlaucht des Fuersten Alexander C. della Torre e Tasso nach Aegypten und auf die Halbinsel Sinai. III. Tenebrionidae (Coleoptera) – Bulletin de la Société Royale Entomologique d'Égypte 18: 17–27; Cairo.

Schaum, H. R. (1859) Beiträge zur europäischen Käferfauna. – Berliner Entomologische Zeitschrift 3: 42–59; Berlin.

Schawaller, W. (1982) Tenebrionidae aus dem Vorderen Orient I (Insecta, Coleoptera). – Stuttgarter Beiträge zur Naturkunde, Serie A (Biologie) 359: 1–14; Stuttgart.

Schawaller, W. (1987) Revision westpaläarktischer Tenebrionidae (Coleoptera). Teil 1. Die Arten der Gattung *Akis* Herbst. – Stuttgarter Beiträge zur Naturkunde, Serie A (Biologie) 403: 1–21; Stuttgart.

Schawaller, W. (1991) The Genus *Diphyrrhynchus* in Arabia, with Description of a New Species (Coleoptera, Tenebrionidae). – Fauna of Saudi Arabia 12: 289–294; Basel.

Schawaller, W. (1993) New and Little Known Tenebrionidae (Coleoptera) from the Arabian Peninsula. – Fauna of Saudi Arabia 13: 102–109; Basel.

Schuster, A. (1922) Tenebrioniden aus der Cyrenaica (Col.). – Memorie della Società Entomologica Italiana 1: 14–23; Genova.

Schuster, A. (1926) Neue Tenebrioniden aus der Cyrenaica II. (Col.). – Bollettino della Società Entomologica Italiana 58: 130–135; Genova.

Schuster, A. (1928) Tenebrionidae. – In: Winkler, A. (1924-1932, ed.): Catalogus Coleopterorum regionis palaearticae: 914–1024; Wien.

Schuster, A. (1934) Tenebrioniden (Col.) vom Toten Meer. – Entomologische Blätter 30 (5): 180–182; Krefeld.

Schuster, A. (1937) Eine Tenebrioniden-Ausbeute von Mesopotamien, Palästina, Syrien und Aegypten. – Koleopterologische Rundschau 23: 44–52; Wien.

Schuster, A. (1938) Neue Tenebrioniden (Col.) von Persisch-Belutschistan. – Koleopterologische Rundschau 24 (3/4): 77–90; Wien.

Schuster, A. & Gebien, H. (1938) Tenebrioniden (Col.) aus Arabien. – Entomologische Blätter 34 (1): 49–62; Krefeld.

Seidlitz, G. von (1890) Fauna Baltica. Die Kaefer (Coleoptera) der deutschen Ostseeprovinzen Russlands. Zweite neu bearbeitete Auflage. Hartungsch Verlagungsdruckerei, [10] + lvi + 192 + 818 pp, 1 pl.; Königsberg.

Seidlitz, G. von (1893) Naturgeschichte der Insecten Deutschlands. Erste Abtheilung Coleoptera. Fünfter Band. Erste Hälfte. 2. Lieferung: 201–400; Berlin.

Seidlitz, G. von (1894) Naturgeschichte der Insecten Deutschlands. Erste Abtheilung Coleoptera. Fünfter Band. Erste Hälfte. 3. Lieferung: 401–608; Berlin.

Semenov-Tjan-Shansky, A. P. & Bogatchev, A. V. (1947) Novyy rod tribu Eurychorini (Col. Tenebrionidae). – Doklady Akademii Nauk Azerbaydzhanskoy SSR 3: 175–177; Baku. [in Russian]

Sénac, H. (1880) Description de quatre Nouvelles espèces appartenant au genre *Pimelia*. – Annales de la Société entomologique de France 10 (5. série): 261–268; Paris.

Sénac, H. (1884) Essai monographique sur le genre *Pimelia* (Fabricius). Première partie. Espèces à tarses postérieurs et intermédiaires comprimés. (1er Division de Solier), i–xix, 1–106; Paris.

Sénac, H. (1887) Essai monographique sur le genre *Pimelia* (Fabricius). Première partie. Espèces à tarses postérieurs et intermédiaires non comprimés. (2e Division de Solier). i–vi, 1–160; Paris.

Sigovini, M., Keppel, E. & Taglipietra, D. (2016) Open Nomenclature in the biodiversity era. – Methods in Ecology and Evolution 7: 1217–1225, London. DOI: 10.1111/2041-210X.12594

Skopin, N. G. (1962) Licinski podsemejstva Pimeliinae. – Trudy vsesoyuznogo nauchnoissledovatel'koso Instituta zashchity rasteniy 7: 191–298; Alma Ata. [in Russian]

Soldati, L. (2009) The Darkling Beetles (Coleoptera: Tenebrionidae) of Qatar. Natura optima dux Foundation. 101 pp + XVII plates; Warszawa.

Solier, A. J. J. (1834) Essai d'une division des Coléoptères Hétéromères, et d'une Monographie de la famille des Collaptèrides. – Annales de la Société entomologique de France 3: 479–636, pl. xii-xv; Paris.

Solier, A. J. J. (1835a) Essai sur les Collaptèrides (suite). – Annales de la Société entomologique de France 4: 249–419, pl. v-ix; Paris.

Solier, A. J. J. (1835b) Essai sur les Collaptèrides (suite). – Annales de la Société entomologique de France 4: 509–572, pl. xiv-xv; Paris.

Solier, A. J. J. (1835c) Prodrome de la famille des xystropides. – Annales de la Société entomologique de France 4: 229–248; Paris.

Solier, A. J. J. (1836a) Essai sur les Collaptèrides (suite). – Annales de la Société entomologique de France 5: 5–200, pl. i-iv; Paris.
Solier, A. J. J. (1836b) Essai sur les Collaptèrides (suite). – Annales de la Société entomologique de France 5: 303–355; Paris.
Solier, A. J. J. (1836c) Essai sur les Collaptèrides (suite). – Annales de la Société entomologique de France 5: 635–684, pl. xxiii-xxiv; Paris.
Solier, A. J. J. (1837) Essai sur les Collaptèrides (suite). – Annales de la Société entomologique de France 6: 151–171, pl. vii; Paris.
Solier, A. J. J. (1838a) Essai sur les Collaptèrides (suite). – Annales de la Société entomologique de France 7: 5–73, pl. i-ii; Paris.
Solier, A. J. J. (1838b) Essai sur les Collaptèrides (suite). – Annales de la Société entomologique de France 7: 159–199, pl. vii; Paris.
Solier, A. J. J. (1844) Essai sur les Collaptérides de la tribu des Molurites. – Memorie della Reale Accademia della Scienze di Torino 6 (2. Ser.): 213–339, pl. i-iv; Torino.
Solier, A. J. J. (1848) Essai sur les Collaptèrides. – Studi entomologici 1: 149–370, pls. IV-XV; Torino.
Solsky, S. M. (1881): Novye i maloizvestnye zhestkokrylye okrain Rossiyskoy imperii i prilezhashchikh stran. – Trudy Russkago Entomologeskago Obscestva 1881-1882, 13: 31–84; St. Petersburg. [in Russian]
Stephens, J. F. (1829) The nomenclature of British insects; being a compendious list of such species as are contained in the Systematic Catalogue of British Insects, and forming a guide to their classification. &c. &c. 68 pp., Baldwin and Cradock; London.
Sturm, J. (1826) Catalog meiner Insecten-Sammlung. Erster Theil. Käfer. – viii, 207 pp., 4 pls.; Nürnberg.
Täckholm, V. L. (1932) Bibliographical notes to the flora of Egypt. – Festschrift till Verner Söderberg; Stockholm: 193–210; Stockholm.
Théront, J. & Hollande, A. (1965) Contribution à l'étude des Coléoptères de la région de Béni-Abbès et de la vallée de la Saoura. – Annales de la Société entomologique de France (N. S.) 1 (4): 851–887; Paris.
Tournier, H. (1868) Coléoptères nouveaux d'Europe, d'Algérie et d'Égypte. – L'Abeille 5: 141–148; Paris.
Tsoar, H. (1995) Desertification in Northern Sinai in the eighteenth century. – Climate Change 29: 429–438;DOI: 10.1007/BF01092427
Uhmann, G., Chikatunov, V. & Pavlíček, T. (2005) Catalogue of the beetles (Coleoptera) in Israel and adjacent areas: 4. Anthicidae: Biocosme Mésogéen, 22 (1): 1–64; Nice.
Vigna Taglianti, A., Audisio, P., A., Biondi, M., Bologna, M., A., Carpaneto, G., M., De Biase, A., Fattorini, S., Piattella, E., Sindaco, R., Venchi, A. & Zapparoli, M. (1999) A proposal for a chorotype classification of the Near East fauna, in the framework of the Western Palearctic region. – Biogeographia - The Journal of Integrative Biogeography 20: 31–59; Siena. DOI: 10.21426/b6110172
Waisel, Y. & Liphschitz, N., 1968. Dendrochronological studies in Israel: II. *Juniperus phoenicea* of North and Central Sinai. – La-Yaaran 18: 2–22, 63–67; Jerusalem.
Walker, F. (1871) List of Coleoptera collected by J. K. Lord, Esc. in Egypt, Arabia and near the African shore of the Red Sea. With characters of the undescribed species. E. W. Janson, 19 pp.; London.
Waltl, J. (1835) Reise durch Tyrol, Oberitalien und Piemont nach dem südlichen Spanien. – Pustet'sche Verlagsbuchhandlung, 247 pp., Passau.
Wendelbo, P. (1961) Studies in Primulaceae. II. An account of *Primula* subgenus *Sphondylia*. – Årbok for Universitetet i Bergen, Mat.-naturv. serie, 11: 5–49; Bergen.
Wollaston, T. V. (1861) On certain Coleoptera from the Island of St. Vincent. – Annals and Magazine of Natural History, including Zoology, Botany, and Geology 3 (7): 197–206, 246–253; London.

Wollaston, T. V. (1864) Catalogue of the Coleopterous Insects of the Canaries in the Collection of the British Museum. 648 pp; London.
Wollaston, T. V. (1867) Coleoptera Hesperidum, being an enumeration of the Coleopterous insects of the Cape Verde Archipelago. John van Voorst, xxxiv+255 pp.; London.
Zahran, M. A. & Willis, A. J. (1992) The vegetation of Egypt. Chapman & Hall, 424 pp; London.
Zalat, S., Gilbert, F., Fadel, H., El-Hawagry, M. S., Saleh, M., Kamel, S. & Gilbert, J. (2008) Biological explorations of Sinai: flora and fauna of Wadi Isla and Hebran, St. Katharine Protectorate, Egypt. – Egyptian Journal of Natural History 5: 6–15; Ismailia. DOI: 10.4314/ejnh.v5i1.70969
Zeini, N. T., Abdel-Hamid, A. M., Soliman, A. S. & Okasha, A. E. (2018) An explanatory study of place-names in Sinai Peninsula, Egypt: a spatial approach. – Annals of GIS 24 (3): 177–194; Berkeley. DOI: 10.1080/19475683.2018.1501759
Zohary, M. (1935) Die phytogeographische Gliederung der Flora der Halbinsel Sinai. – Beihefte zum Botanischen Centralblatt 52 (2): 549–621; Jena.
Zoufal, V. (1892) Uebersicht der mir bekannten Arten der Coleopteren-Gattung *Tenebrio* L. – Wiener Entomologische Zeitung 11 (9): 283–286; Wien.

Taxonomic Index

A

Abantis aenescens Fairmaire, 1892, 72
Accipiter nisus (Linnaeus, 1758), 5
Adelostoma cordatum Solier, 1837, 13, 25, 35
Adelostoma Duponchel, 1827, 25, 35
Adelostoma grande Haag-Rutenberg, 1879, 107
Adelostoma sulcatum crassicorne Peyerimhoff, 1931, 35
Adelostoma sulcatum grandiforme Koch, 1935, 35
Adelostoma sulcatum sulcatum Duponchel, 1827, 14, 25, 35, 83
Adelostomini Solier, 1834, 25, 35
Adelostomoides, Carl, 1991, 107
Adesmia bicarinata bicarinata, 39, 41
Adesmia bicarinata glabrior Schatzmayr & Koch, 1934, 39, 41
Adesmia cancellata abbreviata (Klug, 1830), 10, 15, 23, 38, 39, 40, 83
Adesmia cancellata cancellata (Klug, 1830), 38, 39, 107
Adesmia cancellata clathrata Solier, 1835, 38
Adesmia cancellata hadramautina Kaszab, 1981, 38
Adesmia cancellata latreillei Solier, 38, 39, 96, 107
Adesmia carinata carinata Solier, 1835, 96, 107
Adesmia carinata Solier, 1835, 96
Adesmia convergens Walker, 1871, 44
Adesmia cothurnata cothurnata (Forskål, 1775), 10, 14, 23, 39, 43, 84
Adesmia cothurnata glabrior Schatzmayr & Koch, 1934, 39, 41
Adesmia cothurnata omanensis Kaszab, 1981, 39, 42
Adesmia dilatata dilatata (Klug, 1830), 15, 23, 39, 44, 84
Adesmia dilatata drakei, Crotch, 44
Adesmia dilatata var. *drakei*, 44
Adesmia drakii Crotch, 1872, 44
Adesmia faremontii Lucas, 1844, 108
Adesmia Fischer von Waldheim, 1822, 17, 18, 36
Adesmia gracilis Gory, 107
Adesmia lacunosa (Klug, 1830), 42, 107
Adesmia latreillei Solier, 1835, 38, 39
Adesmia macropus Solier, 1835, 44
Adesmia metallica brozai Ardoin, 1978, 15, 23, 42, 84, 108
Adesmia metallica laevior Ardoin, 1978, 15, 23, 44, 84
Adesmia metallica metallica (Klug, 1830), 108
Adesmia metallica syriaca Baudi di Selve, 1874, 108
Adesmia montana acervata (Klug, 1830), 36, 108
Adesmia montana montana (Klug, 1830), 10, 13, 23, 36, 37, 83, 108, 109
Adesmia montana parallela Miller, 1861, 117
Adesmia perpolita Reitter, 1916, 36
Adesmia sinaitica Crotch, 1872, 38
Adesmia subserrata Chevrolat, 1877, 39
Adesmia tenebrosa Solier, 1835, 108
Adesmiini Lacordaire, 1859, 23, 36
Akidini Billberg, 1820, 24, 44
Akis elevata var. *sculptior* Koch, 1935, 44
Akis Herbst, 1799, 39, 44
Akis reflexa (Fabricius, 1775), 15, 25, 45, 85
Akis spinosa (Linnaeus, 1764), 14, 25, 44, 85
Akis spinosa var. *sculptior* Koch, 1935, 44, 45
Alleculinae Laporte de Castelnau, 1840, 19, 33, 82
Alleculini Laporte de Castelnau, 1840, 33, 82
Allophylax variolosus (Olivier, 1811), 109
Alphitobiini Reitter, 1917, 77
Alphitobius diaperinus (Panzer, 1796), 12, 30, 77, 103
Alphitobius Stephens, 1829, 77
Ammobius Guérin-Méneville, 1844, 71
Ammobius rufus (Lucas, 1846), 13, 17, 32, 71, 100
Ammotrypes Fairmaire, 1879, 75
Amnodeis confluens confluens Miller, 1858, 13, 24, 45, 85
Amnodeis confluens freyi Schuster, 1937, 45
Amnodeis gebieni, Reitter, 1914, 45
Amnodeis Miller, 1858, 45
Anemia aegyptiaca Pic, 1936, 78
Anemia aphodioides Walker, 1871, 79
Anemia asperula Reitter, 1884, 78
Anemia asperula var. *seriesetosa* Baudi di Selve, 1894, 78
Anemia brevicollis, 79
Anemia denticulata Pic, 1923 nec Wollaston, 1867, 78
Anemia fausti Solsky, 1881, 79
Anemia granulata Laporte de Castelnau, 1840, 109
Anemia pharao Reitter, 1897, 79

Anemia pilosa Tournier, 1868, 79
Anemia sardoa, 78
Anemia submetallica Raffray, 1873, 117
Aquila chrysaetos (Linnaeus, 1758), 5
Aquila verreauxii Lesson, 1831, 5
Arthrocnemum macrostachyum (Moric.) C. Koch, 5
Astorthocnemis becvarorum Lillig & Pavlíček, 2002, 15, 17, 21, 49, 87
Astorthocnemis Lillig & Pavlíček, 2002, 49
Atrachyderma, 54

B

Bioplanes syriacus Reiche & Saulcy, 1857, 71
Blaps bifurcata andresi Koch, 1935, 69
Blaps bifurcata bifurcata Solier, 1848, 13, 29, 69, 70, 99
Blaps bifurcata gridellii Pierre, 1961, 70
Blaps bifurcata mirei Gridelli, 1952, 70
Blaps bifurcata strauchi Reiche, 1861, 70
Blaps Fabricius, 1775, 17, 29, 69
Blaps judaeorum Miller, 1861, 13, 29, 70
Blaps laticollis Solier, 1848, 97, 109
Blaps nitens laportei Ardoin, 1973, 13, 29, 70, 99
Blaps pharao Seidlitz, 1893, 70
Blaps polychresta (Forskål, 1775), 13, 29, 70, 99
Blaps sulcata Fabricius, 1775, 70
Blaps sulcata Laporte de Castelnau, 1840 nec Fabricius, 1775, 70
Blaps wiedemanni Solier, 1848, 13, 29, 70
Blaptinae Leach, 1815, 19, 29, 69
Blaptini Leach, 1815, 29, 69
Bufotes sitibundus (Pallas, 1771), 5
Bufotes variabilis, 5
Buteo buteo (Linnaeus, 1758), 5
Buteo rufinus (Cretzschmar, 1829), 5

C

Caedius aegyptiacus Mulsant & Rey, 1859, 117
Calobamon leptoderus Kraatz, 1865, 67
Camelus dromedarius (Linnaeus, 1758), 5
Capra nubiana Cuvier 1825, 5
Caristela megalops Fairmaire, 1894, 82
Cataphronetis apicilaevis Marseul, 1876, 81
Cheirodes asperulus (Reitter, 1884), 14, 31, 78, 104
Cheirodes brevicollis (Wollaston, 1864), 13, 31, 79, 104
Cheirodes Gené, 1839, 17, 19, 31, 78
Cheirodes granulatus (Laporte de Castelnau, 1840), 109
Cheirodes linnmani Ferrer, 2016, 78
Cheirodes pilosus (Tournier, 1868), 14, 32, 79, 104

Cheirodes sardous denticulatus (Wollaston, 1867), 78
Cheirodes sardous sardous Gené, 1839, 11, 15, 32, 78
Cheirodes submetallicus (Raffray, 1873), 117
Ciconia ciconia (Linnaeus, 1758), 5
Ciconia nigra (Linnaeus, 1758), 5
Clitobius grimmi Lillig, 2021, 15, 31, 71, 100
Clitobius Mulsant & Rey, 1859, 31, 71
Clitobius oblongiusculus (Fairmaire, 1875), 14, 31, 71, 72, 101
Clitobius oblongiusculus schneideri Allard, 1882, 62
Clitobius oblongiusculus var. *lineicollis* Fairmaire, 1879, 71, 72
Clitobius ovatus (Erichson, 1843), 109
Colydium castaneum Herbst, 1797, 80
Crataegus sinaica Boiss, 4
Cratopus castaneus Eschscholtz, 1831, 65
Crypticini, Brullé, 1832, 80
Crypticus griseovestris Fairmaire, 1879, 119
Crypticus longulus Reiche & Saulcy, 1857, 118
Crypticus murinus Allard, 1882, 80
Crypticus subpunctatus Walker, 1871, 109
Cteniopodini, Seidlitz, 1896, 33, 82
Cteniopus priesneri luteiceps Pic, 1939, 15, 33, 82
Cteniopus priesneri priesneri Wittmer, 1936, 82
Cteniopus Solier, 1835, 82
Curimosphena heydeni, 57
Curimosphena villosus Haag-Rutenberg, 58
Cyphostethe heydeni (Haag-Rutenberg), 1877, 14, 25, 57, 91
Cyphostethe Marseul, 1867, 17, 57
Cyptus aegyptiacus (Mulsant & Rey, 1859), 117

D

Dailognatha crenata Reiche & Saulcy, 1857, 13, 25, 57, 91
Dailognatha Steven, 1828, 57
Dendarini Mulsant & Rey, 1854, 31, 71
Dendarus calcaroides (Peyerimhoff, 1907), 15, 31, 71, 100
Dendarus Dejean, 1821, 31, 71
Dendarus syriacus (Reiche & Saulcy, 1857), 15, 31, 71, 100
Dermestes navalis Fabricius, 1775, 80
Diaperinae Latreille, 1802, 19, 80
Dichillus alfierii Koch, 1935, 15, 17, 28, 56, 90
Dichillus Jacquelin du Val, 1860, 18, 56
Diphyrrhynchus aenescens (Fairmaire, 1892), 15, 17, 30, 72
Diphyrrhynchus Fairmaire, 1849, 72
Doderoella andreinii Gridelli, 1929, 119
Doderoella cyrenaica Schuster, 1926, 119

Taxonomic Index

Drosochrus costatus aegyptiacus (Gridelli, 1939), 11, 15, 29, 77, 103
Drosochrus costatus costatus Solier, 1848, 77
Drosochrus costatus elegans (Baudi di Selve, 1881), 77
Drosochrus Erichson, 1843, 77
Dymonus tuberculatus Solier, 1844, 56

E

Equus africanus asinus Linnaeus, 1758, 5, 13
Erodiini Billberg, 1820, 23, 45
Erodius bilineatus bilineatus Olivier, 1791, 45, 110
Erodius bilineatus bungemensis Koch, 1937, 110
Erodius bilineatus moccai Koch, 1937, 110
Erodius bilineatus rohlfsi Koch, 1941, 110
Erodius contractus auct. nec Kraatz, 1865, 110
Erodius dejeani var. *lineolatus* Reitter, 1914, 46
Erodius dejeanii duponcheli (Allard), 46
Erodius dejeanii Solier, 1834, 15, 21, 24, 46, 86
Erodius duponchelii Allard, 1865, 15, 24, 46
Erodius ecostatus Crotch, 1872, 111
Erodius Fabricius, 1775, 17, 18, 24, 45
Erodius gibbus cyrenaicus Schuster, 1926, 47
Erodius gibbus gibbus Fabricius, 1775, 15, 24, 47, 86
Erodius glabratus Solier, 1834, 110
Erodius hebraicus Lillig, 1997, 15, 24, 46, 86
Erodius kneuckeri kneuckeri Andres, 1920, 15, 24, 45, 46, 85
Erodius kneuckeri semisculptus Koch, 1935, 46
Erodius maillei auct. nec Solier, 1834, 46, 47
Erodius octocostatus octocostatus Peyrimhoff, 1907, 111
Erodius octocostatus verruculiferus Kazsab, 1982, 111
Erodius opacus Kraatz, 1865, 15, 24, 46
Erodius planus Fabricius, 1775, 69
Erodius puncticollis mirei Pierre 1961, 47
Erodius puncticollis puncticollis Solier, 1834, 47, 110
Erodius puncticollis sinaiticus Crotch, 1872, 10, 15, 26, 46, 47, 48, 86, 110, 111
Erodius puncticollis var. *contractus* Kraatz, 1865, 47
Erodius puncticollis var. *maillei*, 47
Erodius quadrilineatus Kraatz, 1865, 117
Erodius scaber Solier, 1834, 110
Erodius servillei Solier, 1834, 47, 110
Erodius testudinarius Fabricius, 1787, 68
Erodius zophosoides marrakensis Kocher, 1950, 47
Erodius zophosoides nunicus Kocher, 1955, 47
Erodius zophosoides schatzmayri Koch, 1937, 47
Erodius zophosoides subbicostatus Koch, 1937, 47

Erodius zophosoides zophosoides Allard, 1864, 13, 24, 47, 87
Eurycaulus Fairmaire, 1868, 75
Eurycaulus henoni Fairmaire, 1897, 15 32, 75, 102
Eurycaulus peyerimhoffi Reitter, 1904, 75
Eutagenia, 18
Eutagenia arabs, 56

F

Falco tinnunculus (Linnaeus, 1758), 5
Fourtaus brevicornis Pic, 1921, 57

G

Gonocephalum controversum Gridelli, 1948, 118
Gonocephalum costatum rugulosum Küster, 1849, 118
Gonocephalum demaisonis Allard, 1883, 73
Gonocephalum patruele patruele (Erichson, 1843), 14, 31, 72, 101, 111
Gonocephalum patruele turchestanicum Gridelli, 1948, 73
Gonocephalum perplexum (Lucas, 1846), 14, 31, 73, 101
Gonocephalum rusticum (Olivier, 1811), 13, 31, 73, 101
Gonocephalum sericeum (Baudi di Selve, 1875), 111
Gonocephalum setulosum angustum Lindberg, 1950, 73
Gonocephalum setulosum kashgarense (Bates, 1879), 73
Gonocephalum setulosum lindbergi Ferrer, 1993, 73
Gonocephalum setulosum setulosum (Faldermann, 1837), 14, 31, 73, 102
Gonocephalum Solier, 1834, 31, 72
Gonocephalum soricinum faraonicum Ferrer, 2000, 74
Gonocephalum soricinum insidiosum (Fairmaire, 1880), 74
Gonocephalum soricinum soricinum (Reiche & Saulcy, 1857), 14, 31, 73, 74, 102
Gonocephalum strigosum (Reiche, 1847), 111
Gonocephalum tomentosum (Walker, 1871), 111

H

Halocnemum strobilaceum (Pall.) Bieb., 5
Halonomus lineicollis Fairmaire, 1879, 71
Halonomus oblongiusculus Fairmaire, 1875, 71
Halonomus schneideri Allard, 1882, 72
Hedyphanes chikatunovi Nabozhenko & Lillig, 2013, 15, 29, 78, 103
Hedyphanes Fischer, 1820, 78

Hegeterocara arabica Reitter, 1900, 15, 28, 57, 91
Hegeterocara Reitter, 1900, 57
Helopini Latreille, 1802, 29, 78
Helopinus costatus ab. *elegans* Baudi di Selve, 77
Helopinus costatus aegyptiacus Gridelli, 1939, 77
Himatismus heydeni Haag-Rutenberg, 1877, 57
Himatismus perraudierei Marseul, 1867, 67
Himatismus saharensis Chobaut, 1897, 57
Himatismus sublinearis Walker, 1871, 112
Himatismus variegatus Fabricius, 1781, 58, 112
Himatismus villosus (Haag-Rutenberg, 1870), 58, 112
Hionthis Miller, 1861, 58
Hionthis tentyrioides Miller, 1861, 13, 26, 58, 91
Hopatrum sericeum Reiche, 72, 111
Hopatrum sericinum Reiche, 72, 111
Hopatrum tomentosum Walker, 1871, 111
Hyaena hyaena (Linnaeus, 1758), 5
Hyla savignyi Audouin, 1827, 5
Hyoscyamus muticus L., 52

I

Imatismus Dejean 1834, 17, 58
Imatismus sublinearis Walker, 1871, 112
Imatismus variegatus (Fabricius, 1781), 58, 112
Imatismus villosus (Haag-Rutenberg, 1870), 14, 25, 58, 92, 112

J

Juncus rigidus Desf., 5
Juniperus phoenicea L., 3, 4

L

Lagria spec., 19, 112
Lagriinae, 19
Leichenum Dejean, 1834, 77
Leichenum pulchellum Küster, 77
Leichenum pulchellum pulchellum (Lucas, 1846), 77
Leichenum pulchellum pumilum Baudi di Selve, 1876, 14, 30, 77, 103

M

Machlopsis crenatocostata (Redtenbacher, 1868), 13, 25, 36, 83
Machlopsis Pomel, 1871, 36
Melanchrus laevigatus Reiche & Saulcy, 1857, 113
Melancrus pygmaeus Reiche & Saulcy 1857, 63
Melancrus subcostatus Guérin-Méneville, 1862, 113

Melanimonini Seidlitz, 1894, 31, 78
Meostenopa nabathea Peerinhoff, 1907, 112
Mesomorphus longulus (Reiche & Saulcy, 1857), 118
Mesomorphus Miedel, 1880, 74
Mesomorphus murinus murinus, 74
Mesomorphus setosus (Mulsant & Rey, 1853), 15, 30, 74
Mesostena angustata (Fabricius, 1775), 10, 14, 18, 26, 58, 59, 60, 92
Mesostena angustata deserticola Soldati 2009, 59
Mesostena cavatica (Andres, 1926), 112
Mesostena Eschscholtz, 1831, 18, 26, 58
Mesostena laevicollis Solier, 1835, 58, 59
Mesostena longicollis Lucas, 1858, 59
Mesostena nabathaea (Peyerimhoff, 1907), 112
Mesostena oblonga Solier, 1835, 58
Mesostena peyerimhoffi Gridelli, 1938, 15, 26, 59
Mesostena picea picea (Kraatz, 1865), 113
Mesostena picea sinaitica (Schatzmayr & Koch, 1934), 15, 26, 59, 61, 92, 113
Mesostena puncticollis Solier, 1835, 14, 26, 59, 92
Mesostena punctipennis, Solier, 1835, 58
Mesostenopa cavatica Andres, 1926, 112
Mesostenopa gracilis Peyerimhoff, 1907 nec *M. gracilis* Gredler, 1878, 59
Mesostenopa habessinica sinaitica Schatzmayr & Koch, 1934, 59, 61
Mesostenopa picea Klug, 113
Messor intermedius Santschi, 1927, 18
Micipsa batesi Haag-Rutenberg, 1875, 118
Micipsa burtoni Baudi di Selve, 1881, 15, 21, 61, 93
Micipsa cf. *philistina* Reiche & Saulcy, 1857, 13, 19, 61, 93
Micipsa douei distincticornis Peyerimhoff, 1933, 61
Micipsa douei douei Lucas, 1856, 13, 27, 61, 93
Micipsa grandis Kraatz, 1865, 61
Micipsa Lucas, 1855, 27, 61
Micipsa schaumii Kraatz, 1865, 15, 27, 62, 93
Microtelus asiaticus Solier, 1838, 113
Microtelus careniceps binodiceps Reitter, 1907, 15, 28, 56
Microtelus careniceps careniceps Reiche & Saulcy, 1857, 56
Microtelus careniceps sinaiticus Koch, 1935, 56
Microtelus lethierryi Reiche, 1860, 56, 118
Microtelus Solier, 1838, 56
Microtelus torretassoi Koch, 1934, 118
Mitotagenia aegyptiaca Koch, 1941, 13, 28, 56
Mitotagenia arabs (Baudi di Selve, 1881), 57
Mitotagenia arabs auct. nec (Baudi di Selve, 1881), 56
Mitotagenia Reitter, 1916, 56

Taxonomic Index

Mycetocharina longipennis Pic, 1923, 82
Mycetocharina megalops megalops (Fairmaire, 1894), 14, 33, 82
Mycetocharina megalops pallidior Pic, 1938, 82
Mycetocharina Seidlitz, 1890, 82
Myrmechixenus, 18

N

Neocaedius aegyptiacus (Mulsant & Rey, 1859), 117

O

Ocnera andresi Gridelli, 1928, 53
Ocnera genei Solier, 1836, 53
Ocnera gomorrhana (Reiche & Saulcy, 1857), 54
Ocnera habelmanni Kraatz, 1865, 55
Ocnera hispida (Forskål, 1775), 54
Ocnera hispida latreillei Solier, 1836, 54
Ocnera parvicollis Baudi di Selve, 1875, 55
Ocnera philistina (Reiche & Saulcy, 1857), 55
Omophlus Dejean, 1834, 82
Omophlus gracilipes Kirsch, 1869, 13, 33, 82
Onychognathus tristrami (Scatler, 1858), 5
Opatrini Brullé, 1832, 31, 71
Opatrinus setosus Mulsant & Rey, 1853, 74
Opatroides Brullé, 1832, 74
Opatroides punctulatus lilligi Ferrer, 2005, 75
Opatroides punctulatus parvulus (Faldermann, 1837), 75
Opatroides punctulatus punctulatus Brullé, 1832, 10, 14, 30, 74, 75, 76
Opatroides subcylindricus Ménétriés, 1849, 75
Opatrum armatum Waltl, 1835, 120
Opatrum hirtulum Baudi di Selve, 1875, 75
Opatrum murinum Baudi di Selve, 1876, 74
Opatrum orientale Fabricius, 1775, 75
Opatrum ovatum Erichson, 1843, 109
Opatrum patruele Erichson, 1843, 72
Opatrum perplexum Lucas, 1846, 73
Opatrum rugulosum Küster, 1849, 118
Opatrum rusticum Olivier, 1811, 73
Opatrum sericeum Baudi di Selve, 111
Opatrum setulosum Faldermann, 1837, 73
Opatrum soricinum Reiche & Saulcy, 1857, 73
Opatrum strigosum Reiche, 1847, 111
Opatrum variolosus Olivier, 1811, 109
Oterophloeus alveatus alveatus Peyerimhoff, 1931, 62, 94
Oterophloeus alveatus peyerimhoffi Koch, 1935, 15, 26, 62
Oterophloeus Desbrochers des Loges, 1881, 26, 62
Oterophloeus haagii haagii (Kraatz, 1865), 62
Oterophloeus haagii orientalis Koch, 1935, 14, 15, 26, 62, 94

Oxycara aelaniticum Peyerimhoff, 113
Oxycara ardoini Kaszab, 1979, 15, 28, 62, 94
Oxycara breviusculum Fairmaire, 1892, 63
Oxycara laevigatum (Reiche & Saulcy, 1857), 113
Oxycara peyerimhoffi Lillig, 2001, 10, 15, 28, 63, 64, 95, 114
Oxycara producta Peyerimhoff, 1907, 54
Oxycara productum Peyerimhoff, 1907, 15, 62
Oxycara pygmaeum (Reiche & Saulcy, 1857), 13, 28, 63, 94
Oxycara Solier, 1835, 28, 62
Oxycara subcostatum (Guérin-Méneville, 1862), 63, 113

P

Pachyscelis rotundata Kraatz, 1865, 4
Palorus, 18
Panthera pardus (Linnaeus, 1758), 5
Pedinini Eschscholtz, 1829, 29, 30, 77
Pelophylax bedriagae (Camerano, 1882), 5
Penthicus punctulatus (Brullé, 1832), 74
Phaleria acuminata syriaca Reitter, 1916, 119
Phaleria aegyptiaca Seidlitz, 1894, 81
Phaleria Latreille, 1802, 81
Phaleria munda Walker, 1871, 81
Phaleria prolixa Fairmaire, 1869, 14, 17, 32, 81, 105
Phaleria syriaca Reitter, 1916, 119
Phaleriini Blanchard, 1845, 81
Philhammus aharonii (Reitter, 1910), 119
Phoenix dactylifera L., 5
Phragmites australis (Cav.) Trin. ex Steud., 5
Phtora angusta (Wollaston, 1861), 81
Phtora apicilaevis (Marseul, 1876), 11, 15, 32, 81
Phtora Germar, 1836, 32, 81
Phtora subclavata (Wollaston, 1861), 11, 15, 32, 81, 105
Phylax calcaroides Peyerimhoff, 1907, 61
Pimelia abbreviata Klug, 1830, 38, 39
Pimelia acervata Klug, 1830, 108
Pimelia aculeata Klug, 1830, 49
Pimelia alternata Klug, 1830, 49
Pimelia angulata angulata Fabricius, 1775, 15, 20, 49, 87
Pimelia angulata angulosa A. G. Olivier, 1795, 49, 50
Pimelia angulata berlandi Ardoin, 1963, 49
Pimelia angulata nilotica Sénac, 1884, 49, 114
Pimelia angulata sinaitica Schatzmayr & Koch, 1934, 49
Pimelia angulata syriaca Sénac, 1884, 49
Pimelia angulata tschadensis Koch, 1940, 49
Pimelia angustata angustata Fabricius, 1775, 58
Pimelia arabica arabica Klug, 1830, 50

Pimelia arabica edomita Koch, 1940, 15, 20, 50, 87
Pimelia arabica emiri Kaszab, 1982, 50
Pimelia arabica omanica Kaszab, 1982, 50
Pimelia asperata Klug, 1830, 115
Pimelia bajula Klug, 1830, 114, 115
Pimelia barthelemyi Solier, 1836, 13, 21, 50, 88
Pimelia bicarinata Klug, 1830, 39
Pimelia bottae Sénac, 1887, 14, 20, 51
Pimelia bottai Sénac, 51
Pimelia cancellata Klug, 1830, 107
Pimelia canescens arabicola Schatzmayr & Koch, 1934, 51
Pimelia canescens canescens Klug, 1830, 13, 20, 51, 88
Pimelia coronata Olivier, 1795, 52
Pimelia derasa barcana Koch, 1941, 114
Pimelia derasa derasa Klug, 1830, 114
Pimelia derasa derasoides Schuster, 1922, 119
Pimelia dilatata Klug, 1830, 44
Pimelia Fabricius, 1775, 17, 49
Pimelia grandis grandis Klug, 1830, 17
Pimelia granulata tenuicornis Solier, 1836, 114
Pimelia granulata Solier, 1836, 102
Pimelia hirtella Sénac, 1887, 13, 20, 51
Pimelia interpunctata andreinii, 119
Pimelia interpunctata cyrenaica, 119
Pimelia interpunctata Klug, 1830, 119
Pimelia irrorata Solier, 1836, 51
Pimelia lacunosa Klug, 1830, 107
Pimelia metallica Klug, 1830, 108
Pimelia mittrei Solier, 1836, 15, 20, 51, 88
Pimelia montana Klug, 1830, 36
Pimelia nilotica Sénac, 1884, 114
Pimelia nilotica var. *aegyptiaca* Andres, 1929, 49, 114
Pimelia orbiculata Fabricius 1775, 115
Pimelia reflexa Fabricius, 1775, 45
Pimelia sericea Olivier, 1795, 115
Pimelia subquadrata subquadrata Sturm, 1826, 15, 21, 51, 88
Pimelia tenebrosa Solier, 1835, 108
Pimelia tenuicornis, Solier, 1836, 114
Pimelia theveneti Sénac, 1880, 15, 20, 52
Pimelia thomasi Blair, 1931, 50
Pimeliinae Latreille, 1802, 19, 20, 35
Pimeliini Latreille, 1802, 17, 19, 20, 49
Pistacia khinjuk Stock, 4
Primula boveana Decne, 4
Prionotheca coronata coronata (Olivier, 1795), 14, 21, 52, 89
Prionotheca coronata freyi Bytinski-Salz, 1955, 52
Prionotheca coronata ovalis Ancey, 1881, 52
Prionotheca Dejean, 1834, 52
Procavia cappensis (Pallas, 1766), 5
Prochoma audouini Solier, 1835, 14, 25, 63, 95

Prochoma Solier, 1835, 63
Proscheimus arabicus Desbrochers des Loges, 1881, 13, 32, 72
Proscheimus Desbrochers des Loges, 1881, 72
Pseudanemia brevicollis Wollaston 1864, 79
Pseudophylothes sinaicus Nakamura, 1975, 5, 6
Pseudoseriscius griseovestris (Fairmaire, 1879), 119
Pseudoseriscius maculosus maculosus Fairmaire, 1870, 80
Pseudoseriscius maculosus murinus (Allard, 1882), 15, 32, 80
Pseudostene subclavata Wollaston, 1861, 81
Psilachnopus aharonii Reitter, 1910, 119
Pterolasia Solier, 1836, 52
Pterolasia squalida Solier, 1836, 14, 20, 52, 89
Ptychadena nilotica (Seetzen, 1855), 5

S

Satyrium jebelia (Nakamura, 1975), 5
Scaurini Billberg, 1820, 29, 79
Scaurus aegyptiacus Solier, 1838, 14, 30, 79
Scaurus barbarus Solier, 1838, 115
Scaurus Fabricius, 1775, 29, 79
Scaurus giganteus Küster, 1848, 79
Scaurus puncticollis v. *rugicollis* Reitter, 1914, 80
Scaurus rugicollis Reitter, 1914, 15, 30, 80
Scaurus tristis Olivier, 1795, 115
Scelosodis castaneus castaneus (Eschscholtz, 1831), 14, 27, 65, 95
Scelosodis castaneus soudanicus Girard & Pierre, 1965, 65
Scelosodis Solier, 1835, 65
Schweinfurthia alfierii Schatzmayr & Koch, 1934, 15, 27, 65
Schweinfurthia Andres, 1922, 27, 65
Schweinfurthia sinaitica Andres, 1922, 15, 27, 65
Scleron sulcatum lomii Gridelli 1953, 115
Scleron sulcatum saharense Peyerimhoff, 1931, 115
Scleron sulcatum scotti Gridelli, 1953, 115
Scleron sulcatum sulcatum Baudi di Selve, 1876, 115
Scleronimon Reitter, 1904, 75
Scleropatroides hirtulus (Baudi di Selve, 1875), 13, 19, 32, 75, 102
Scleropatroides Löbl & Merkl, 2003, 75
Sclerophrys regularis (Reuss, 1833), 5
Sclerum armatum (Waltl, 1835), 120
Sclerum Dejean, 1834, 75
Sclerum multistriatum (Forskål, 1775), 120
Sclerum orientale orientale (Fabricius, 1775), 14, 32, 75
Sclerum orientale yemense Kaszab, 1972, 77
Sepidiini Eschscholtz, 1829, 28, 55

Sepidium abiram Crotch, 1872, 55
Sepidium dathan Crotch, 1872, 13, 28, 55, 90
Sepidium Fabricius, 1775, 28, 55
Sepidium korah Crotch, 1872, 55
Sepidium tricuspidatum tricuspidatum Fabricius, 1775, 13, 28, 55, 90
Silpha multistriata Forskål, 1775, 120
Steira crenato-costata Redtenbacher, 1868, 36
Stenosini Schaum, 1859 (1934), 28, 56
Stenosis affinis (Solier, 1838), 15, 28, 57
Stenosis Herbst, 1799, 57
Storthocnemis Karsch, 1881, 49
Symphoxycara Koch, 1943, 113

T

Tagenia affinis Solier, 1838, 57
Tenebrio cothurnatus Forskål, 1775, 39
Tenebrio diaperinus Panzer, 1796, 77
Tenebrio hispidus Forskål, 1775, 54
Tenebrio polychrestus (Forskål, 1775), 70
Tenebrio spinosa Linnaeus, 1764, 44, 45
Tenebrio syriacus Zoufal, 1892, 120
Tenebrio variegatus Fabricius, 1781, 112
Tenebrioninae Latreille, 1802, 19, 29, 77
Tentyria aegyptiaca Solier, 1835, 66
Tentyria discicollis Reiche & Saulcy, 1857, 15, 27, 65, 95
Tentyria giraffa Allard, 1883, 66
Tentyria glabra auct. nec (Fabricius, 1775), 66
Tentyria laticollis Kraatz, 1865, 65
Tentyria Latreille, 1802, 27, 65
Tentyria longicollis (Lucas, 1855), 116
Tentyria punctatostriata Solier, 1835, 13, 27, 65
Tentyria sinaitica Peyerimhoff, 1907, 15, 17, 27, 66, 96
Tentyriini Eschscholtz, 1831, 25, 57
Tentyrina boehmi subsulcata Reiche & Saulcy, 1857, 66
Tentyrina haagi (Kraatz, 1865), 62
Tentyrina longicollis (Lucas, 1855), 116
Tentyrina orbiculata orbiculata (Fabricius, 1775), 66, 115
Tentyrina orbiculata subsulcata (Reiche & Saulcy, 1857), 13, 27, 66, 96, 115
Tentyrina palmeri palmeri (Crotch, 1872),14, 27, 66, 67, 96
Tentyrina palmeri thomasi (Blair, 1931), 67
Tentyrina Reitter, 1900, 66
Thraustocolus Kraatz, 1865, 67
Thraustocolus leptoderus (Kraatz, 1865), 15, 27, 67, 96
Thriptera crinita crinita (Klug, 1830), 14, 21, 52
Thriptera crinita gilleti Pierre, 1979, 53
Thriptera guyoti Andres, 1920, 53
Thriptera heydeni Reitter, 1893, 53

Thriptera kraatzi Haag-Rutenberg, 1876, 14, 21, 53, 89
Thriptera lanata Peyerimhoff, 1907, 52
Thriptera Solier, 1836, 21, 52
Thriptera sorocula Reitter, 1894, 53
Thriptera varvasi pilipes Kraatz, 1865, 15, 21 53
Thriptera varvasi varvasi Solier, 1836, 53
Thymus decussatus (Bentham, 1834), 6
Trachycelis rufus (Lucas, 1846), 71
Trachyderma andresi (Gridelli, 1928), 15, 22, 53
Trachyderma genei Solier, 1836, 13, 22, 47, 53
Trachyderma hispidum (Forskål, 1775), 14, 21, 54, 89
Trachyderma hispidum gomorrhanum Reiche & Saulcy, 1857, 54
Trachyderma hispidum latreillei Solier, 1836, 54
Trachyderma Latreille, 1828, 21, 53
Trachyderma philistina Reiche & Saulcy, 1857, 55
Trachyderma philistinum Reiche & Saulcy, 1857, 14, 22, 48, 55, 90
Trachyscelini Blanchard, 1845, 82
Trachyscelis Latreille, 1809, 82
Trachyscelis rufus Lucas, 1846, 71
Trachyscelis tenuestriatus Fairmaire, 1886, 15, 32, 82, 105
Triboliini Gistel, 1848, 30, 80
Tribolium castaneum (Herbst, 1797), 12, 13, 30, 80, 104
Tribolium confusum Jacquelin du Val, 1861, 12, 13, 30, 80, 105
Tribolium MacLeay, 1825, 30, 80
Tribolium navale Fabricius, 80
Trichosphaena perraudierei perraudierei (Marseul, 1876), 14, 25, 67, 97
Trichosphaena perraudieri peyerimhoffi Gridelli, 1950, 67
Trichosphaena Reitter, 1916, 67

V

Vachellia tortilis subsp. *raddiana* (Savi) Kyal. & Boatwr., 4
Vachellia tortilis subsp. *tortilis* (Forssk.) Galasso & Banf., 4
Vieta Laporte de Castelnau, 1840, 56
Vieta tuberculata (Solier, 1844), 14, 28, 56

Z

Zarudnionymus Semenov-Tjan-Shansky & Bogatchev, 1947, 107
Zilla spinosa (L.) Prantl, 52
Ziziphus spina-christi, 4
Zophosini Solier, 1834, 22, 67
Zophosis aelanitica Peyerimhoff, 1907, 67, 113

Zophosis bicarinata cf. *ghilianii*, Deyrolle, 1867 11, 15, 22, 68, 97, 116
Zophosis bicarinata quadricostata Solier, 1834, 116
Zophosis complanata Solier, 1834, 13, 22, 68, 98
Zophosis depressipennis Lucas, 1858, 69
Zophosis farinosa Blair, 1914, 14, 22, 68
Zophosis Latreille, 1802, 18, 67
Zophosis lethierryi Deyrolle, 1867, 13, 23, 68, 98
Zophosis maillei Solier, 1834, 67, 68
Zophosis nigroaenea Deyrolle, 1867, 13, 23, 67, 97, 113
Zophosis personata Erichson, 1841, 14, 23, 69, 98

Zophosis pharaonis pharaonis Reitter, 1916, 13, 22, 69, 98
Zophosis pharaonis simplex Kaszab, 1981, 69
Zophosis plana (Fabricius, 1775), 14, 22, 69, 99
Zophosis punctata alborana Baudi di Selve, 1883, 68
Zophosis punctata punctata Brullé, 1832, 13, 23, 67, 68, 97
Zophosis punctata pygmaea Solier, 1834, 68
Zophosis rotundata Deyrolle, 1867, 69
Zophosis rotundata Ménétries, 69
Zophosis testudinaria Solier, 1834, nec *Erodius testudinarius* Fabricius, 1787, 68